Infoterm Series 5

Proceedings of the First International Conference

Vienna
2 and 3 April 1979
convened
by Infoterm

Terminological Data Banks

K·G·Saur
München · New York · London · Paris

General editor:

International Information Centre for Terminology
Centre International d'Information pour la Terminologie
Международный информационный центр по терминологии
Internationales Informationszentrum für Terminologie

Infoterm is affiliated to ON (Austrian Standards Institute)
Leopoldsgasse 4 · A-1021 Wien 2 · Austria

Infoterm, established in 1971, is sponsored by Unesco within the framework of the UNISIST programme. It works in liaison with the Technical Committee 37 "Terminology (principles and co-ordination)" of the International Organization for Standardization (ISO).

Infoterm receives financial support from
Bundesministerium für Bauten und Technik,
Bundeskammer der gewerblichen Wirtschaft,
Österreichisches Normungsinstitut

Issue editor:

Christian Galinski (Infoterm)

CIP-Kurztitelaufnahme der Deutschen Bibliothek

Terminological data banks : proceedings of the 1. internat. conference / convened, Vienna 2 and 3 April 1979, by Infoterm. – München, New York, London, Paris : Saur, 1980.
(Infoterm series ; 5)
ISBN 3-598-21365-4

NE: Internationales Informationszentrum für Terminologie ⟨Wien⟩

Druck/Binden: Hain-Druck KG, Meisenheim/Glan
Printed in the Federal Republic of Germany
ISBN 3-598-21365-4

TABLE OF CONTENTS

1 INTRODUCTION

The pressing need for co-ordination of terminological activities carried out throughout the world has become increasingly apparent since the sixties.

In order to meet this need Infoterm was established in 1971 with the assistance of Unesco within the framework of UNISIST (now General Information Programme of Unesco) to function as focal point of co-ordination.
Infoterm is affiliated to the Austrian Standards Institute.

In the last decade computer aided methods have been introduced in terminology work, which gave rise to the establishment of terminological data bases and banks in different parts of the world. Terminological data banks are important reference tools for coding and decoding of information. They will be integrated more and more in information systems and networks.

In the near future it is to be expected that terminological data of concepts and of concept structures of most subject fields will be recorded in machine-readable form, will be updated and stored in terminological data banks. These data will also be offered to information systems and networks. This work will require close co-operation between subject specialists, terminological data bank experts, linguists and information specialists.
Considering the great number of subject fields and their interdisciplinary character as well as the great number of languages to be covered and the rapid increase of knowledge caused by progress in all fields of human life, terminological data banks have to meet manifold specific user needs. This results in different data structure and organization of terminological data banks.

Since problems exist which are common to all these banks this First International Conference on Terminological Data Banks was convened by Infoterm under the auspices and with the assistance of Unesco in order to explore the possibilities of a better international co-operation in this field.

It is hoped that such meetings will be held on a regular basis and will represent the forum for discussion of such problems and future developments.

Vienna, February 1980

Prof. Helmut Felber
Director of Infoterm

2 LIST OF PARTICIPANTS

2.1 Alphabetical list of participants with addresses

Mr. L. ALGOTSSON; Director, Tekniska nomenklaturcentralen (TNC), Box 5243, 102 45 Stockholm, Sweden

Dr. Ali M. AL-KASIMI; Bureau of Coordination of Arabization in the Arab World, 10, rue d'Angola, P.O. Box 290, Rabat, Morocco

Dr. A. ANDREYEWSKY; Technical Translation Systems, LTD, Brookdale Road, R.D. 4, Hahopac, New York 10541, USA (for Carnegie-Mellon University)

Dr. D. ARNOLD; Chief, Documentation and Terminology Section, United Nations (UN), Room 2494, New York 10017, USA

Mr. P. AUGER; Director, Office de la langue française (OLF), 700 Bd. Saint-Cyrille est, Québec, G1R 5A9, Canada

Mr. J.A. BACHRACH; Head of Translation Division, Commission of the European Communities (CEC), Bâtiment Jean Monnet, Plateau du Kirchberg, Boîte postale 1907, Luxembourg - Grand Duché

Mr. J. BAUDOT; Director, Centre de Calcul, Université de Montréal, C.P. 6128, Montréal 101, H3C 3J7, Canada

Mr. A. BOTHE; Chef du service de terminologie et de documentation, AVT/TD, Ministère des Affaires Etrangères, Casuariestraat 16, La Haye, Pay-Bas.

Mr. K.H. BRINKMANN; Siemens AG, Sprachendienst, Postfach 70 00 73, D-8000 München 70, FRG

Dr. F. BRUNS; c/o Bundesminister der Verteidigung, - S III 4 -, Postfach 1328, D-5300 Bonn 1, FRG

Mr. N. CASSAR; Documentation and Terminology Section, United Nations (UN), Room 2494c, New York 10017, USA

Prof. A. CLAS; Director, Departement de linguistique, Université de Montréal, Montréal 101, C.P. 6128, H3C 3J7, Canada

Mr. B. de BESSE; Association française de terminologie (AFTERM), Haut Comité de la langue française, 32, rue de babylone, 75007 Paris, France

Dr. M.A. DOVBENKO; Director, Vsesojuznyj naučno-issledovatel'skij institut tehničeskoj informacii, klassifikacii i kodirovanija Gosstandarta SSSR (VNIIKI), ul. Ščuseva 4, 103001, Moskva, K-1, USSR

MinRat. Dipl.-Ing. E. DÜBELL; Bundesministerium für Bauten und Technik, Abt. IV/2, Stubenring 1, A-1010 Wien, Austria

Ms. D. EICHFELDER; Chief, Reference and Terminology Unit, United Nations Industrial Development Organization (UNIDO), P.O. Box 707, Rathausplatz 2, A-1011 Wien, Austria

Mr. G. ENGEL; Handelshøjskolen i København, Terminologi - afdelingen, Fabrikvej 7, 2000 København F., Denmark

Prof. H. FELBER; Director, Infoterm, Postfach 130, Leopoldsgasse 4, A-1021 Wien, Austria

Mr. Ch. GALINSKI; Infoterm, Postfach 130, Leopoldsgasse 4, A-1021 Wien, Austria

Mr. J. GOETSCHALCKX; Chef du Bureau de terminologie, Commission des Communautés Européennes (CCE), Bâtiment Jean Monnet, B.A. 1/074, Plateau du Kirchberg, Boîte postale 1907, Luxembourg - Grand Duché

Prof. Dr. R. GOFFIN; Bureau de terminologie, IX/D/3, Commission des Communautés Européennes (CCE), 23-27, avenue de la Joyeuse Entrée, B.P. 1049, Bruxelles, Belgique

Ms. I. GORM-HANSEN; Handelshøjskolen i København, Terminologi - afdelingen, Fabrikvej 7, 2000 København F., Denmark

Mr. J.F. GREGOIRE; Faculté des Arts et des Sciences, Université de Montréal, C.P. 6128, Montréal 101, H3C 3J7, Canada

Prof. K. HEGGSTAD; The Norwegian Termbank Nordisk institutt, PDS, Harald Harfagresgt. 29, 5014 Bergen-Universitet, Norway

Dr. M.T. KHAFAGY; Director, Department of Documentation and Information, Arab Educational, Cultural and Scientific Organization (ALECSO), 109 Tahrir Street, Dokki-Cairo, Giza, The Arab Republic of Egypt

Mr. P. KIROUAC; Banque de terminologie du Québec, Office de la langue française (OLF), 700 Bd. Saint-Cyrille est, G1R 5A9, Québec, Canada

Dr. H. KOWALSKI; Chef du Bureau de terminologie, Commission des Communautés Européennes (CCE), 200, rue de la Loi, 1049 Bruxelles, Belgique

Mr. F. KROLLMANN; Leitender Regierungsdirektor, Bundessprachenamt, Horbeller Straße, D-503 Hürth, FRG

Ms. M. KROMMER-BENZ, M.A.; Infoterm, Postfach 130, Leopoldsgasse 4, A-1021 Wien, Austria

Mr. A. LANDRY; Acting Director General, Terminology and Documentation Directorate, Secretary of State, Translation Bureau, Ottawa K1A OM5, Canada

Dr. A. LANE; World Information Centre for legal, economic and administrative terminology (LEAT), Forstenriederallee 4a, D-8000 München 71, FRG

Ms. I.S.F. LAURENT; Normaterm, Association française de normalisation (AFNOR), Tour Europe, Cedex 7, 92080 Paris la Défense, France

Mr. P. LE QUELLEC; Sous secrétaire d'Etat adjoint, Bureau des traductions, Secrétariat d'Etat, Ottawa, K1A OM5, Canada

Dr. W. LÖHNER; Division of the General Information Programme, Unesco, 7, Place de Fontenoy, 75700 Paris, France

Mr. H. LUTZ; Assistant Director of marketing services, Brown Boveri & Cie Aktiengesellschaft (BBC), CH-5401 Baden, Switzerland

Mag. A. MANU; Infoterm, Postfach 130, Leopoldsgasse 4, A-1021 Wien, Austria

Mr. P. MARTIN; Project manager, Terminological data bank, Brown Boveri & Cie Aktiengesellschaft (BBC), CH-5401 Baden, Switzerland

Mr. P.R. MARTIN; Secrétariat du Comité d'Etudes n°.1, Commission Electrotechnique Internationale (CEI), 1, avenue du Général-de-Gaulle, 92140 Clamart, France

Dr. D. McCracken; Computer Science Department, Carnegie-Mellon University, Schenley Park, Pittsburgh, Pennsylvania 15213, USA

Mr. M. PALNIKOV; European Coordination Centre for Research and Documentation in Social Sciences, Grünangergasse 2, A-1010 Wien, Austria

Mr. P. PORT; M.A.; Eschenheimer Anlage 20, D-6000 Frankfurt/M. 1, FRG

Prof. D. RAVENTOS de CASTRO; Director, Proyecti Banco de Terminologia, Departamento de Idiomas, Universidad Simon Bolivar, P.O. Box 1512, Caracas 101, Venezuela

Mr. L. ROLLING; DG 13B, Commission des Communautés Européennes (CCE), Jean Monnet Centre Européen, B 4024, Kirchberg, Luxembourg - Grand Duché

Prof. Dr. G. RONDEAU; Département de langues et linguistique, Faculté des Lettres, Université Laval, Québec, G1K 7P4, Canada

Prof. J.C. SAGER; Head of Department of European Studies and Modern Languages U.M.I.S.T., P.O. Box 88, Manchester M60 1QD, United Kingdom

Mr. W.H.U. SCHEWE; DIN Deutsches Institut für Normung e.V. Normenausschuß Terminologie (NAT), Postfach 1107, Burggrafenstr. 4-10, D-1000 Berlin 30, FRG

Ms. H. SUONUUTI; Centralen för Teknisk Terminologi, Tekniikan Sanastokeskus r.y. (TSK), Liisankatu 16 A 8, 00170 Helsinki 17, Finland

Mr. A. TCHIRIKOFF; Chief of Terminology and Technical Documentation Section United Nations Office at Geneva, Palais de Nations, Room E 5102, CH-1211 Geneva 10, Switzerland

Prof. Dr. E. WEIS; Institut für Romanische Sprachen, Wirtschaftsuniversität, Franz Klein Gasse 1, A-1190 Wien, Austria

Prof. Dr. A. ZAMPOLLI; C N U C E, Via Santa Maria, 36, 56100 - Pisa, Italy

Mr. Iu. ZDOROVOV; Vsesojuznyj naučno-issledovatel'skij institut tehničeskoj informacii, klassifikacii i kodirovanija Gosstandarta SSSR (VNIIKI), 103001, Moskva, K-1, ul. Ščuseva 4, USSR

2.2 List of participants classified by countries following UDC

UDC	COUNTRIES
(410)	**United Kingdom** Mr. Sager (UMIST) (term bank in planning)
(430.1)	**Federal Republic of Germany** Mr. Brinkmann (Siemens) (term bank in operation) Mr. Bruns (BM-Verteidigung) Mr. Krollmann (Bundessprachenamt) (term bank in operation) Mr. Lane (LEAT) Mr. Port (affiliation DGD-KTS) (term bank in development) Mr. Schewe (DIN) (term bank in development)
(436)	**Austria** Mr. Dübell (BMf.BuT) Mr. Weis (Wirtschaftsuniversität Wien)
(44)	**France** Mr.de Bessé (AFTERM) (term bank in planning) Ms. Laurent (Normaterm, AFNOR) (term bank in operation)
(45)	**Italy** Mr. Zampolli (University of Pisa)
(47 + 57)	**USSR** Mr. Dovbenko (VNIIKI) (term bank in operation) Mr. Zdorovov (VNIIKI)
(480)	**Finland** Ms. Suonuuti (TSK)
(481)	**Norway** Mr. Heggstad (University Bergen) (term bank in operation)
(485)	**Sweden** Mr. Algotsson (TNC) (term bank in operation)
(489)	**Denmark** Mr. Engel (Handelshøjskolen) (term bank in development) Ms. Gorm-Hansen (Handelshøjskolen) (term bank in development)
(492)	**Netherlands** Mr. Bothe (AVT/TD)
(494)	**Switzerland** Mr. Lutz (BBC) Mr. Martin (BBC) (term bank in operation)

UDC COUNTRIES

(620) Egypt

Mr. Khafagy (ALECSO)

(64) Morocco

Mr. Al-Kasimi (ALECSO)

(71) Canada

Mr. Auger (OLF)
Mr. Baudot (Université de Montréal)
Mr. Clas (Université de Montréal)
Mr. Grégoire (Université de Montréal)
Mr. Kirouac (OLF, Banque de terminologie) (term bank in operation)
Mr. Landry (DGTD, Secrétariat d'Etat) (term bank in operation)
Mr. Le Quellec (Secrétariat d'Etat)
Mr. Rondeau (Université Laval)

(73) USA

Mr. Andreyewsky (Carnegie-Mellon University)
Mr. McCracken (Carnegie-Mellon University) (term bank in development)

(87) Venezuela

Ms. Raventos de Castro (Universidad Caracas) (term bank in development)

2.3 International and European Organizations and their representatives at the Conference

Commission of European Communities (CEC)

Mr. Bachrach (Luxembourg) (term bank in operation)
Mr. Goetschalckx (Luxembourg) (term bank in operation)
Mr. Goffin (Brussels) (term bank in operation)
Mr. Kowalski (Brussels) (term bank in operation)
Mr. Rolling (Luxembourg)

European Coordination Centre for Research and Documentation (Vienna)

Mr. Palnikov

Infoterm/ON (Vienna)

Mr. Felber
Mr. Galinski
Ms. Krommer-Benz
Mr. Manu

International Electrotechnical Commission (IEC)

Mr. Martin (IEC/TC 1) (term bank in planning)

International Institute for Legal and Administration Terminology (LEAT)

Mr. Lane

Unesco

Mr. Löhner (General Information Programme)
Ms. Krommer-Benz (representing INTERCONCEPT) (term bank in planning)

UNIDO (Vienna)

Ms. Eichfelder

United Nations (Geneva)

Mr. Tschirikoff

United Nations (New York)

Mr. Arnold (term bank in development)
Mr. Cassar (term bank in development)

3 LIST OF PAPERS AND REPORTS DISTRIBUTED FOR THE CONFERENCE

Invited papers

RONDEAU, G. Typologie des banques de terminologie

SUNDSTRÖM, E. Transfer modalities of terminological data

BRINKMANN, K.-H. Quality criteria for the exchange of terminological data

Invited reports

FELBER,H. Report on TermNet

BOTHE, A. The systematic classification of terminology / La classification systématique des stocks terminologiques

KROMMER-BENZ, M. State-of-the-art report: Project "Vocabulary of terminological data banks"

Other reports and papers

AL-KASIMI, A. Towards a central terminological data bank in the Arab world

DOVBENKO, M.A. Content and presentation of data elements for terminological data interchange

DOVBENKO, M.A. Gosstandart activities in terminology and terminological data banks interaction

FAO. International co-operation between terminological data banks. Views and expectations of the organizations of the United Nations system. (Distributed with the Report Doc. TermNet 4 (9)-79)

GOFFIN, R. The vocabulary of terminological data banks

KHAFAGY, M.T. ALECSO and its efforts at unifying arabic terminology and the standardization of its application

KROMMER-BENZ, M. TermNet and INTERCONCEPT

RIGGS, F. Preliminary suggestion with reference to INTERCONCEPT

RONDEAU, G. The terminology bank of Canada

SCHEWE, W.H.U. Recent experiences with the establishment of a terminological data bank within DIN

SECRETARY OF STATE. TERMINOLOGY DIRECTORATE. The Canadian government terminology bank

SECRETARY OF STATE. TERMINOLOGY DIRECTORATE. Subject-field descriptions. Working document of classification group

SECRETARIAT D'ETAT. DIRECTION DE LA TERMINOLOGIE. Description des domaines. Document de travail à l'usage du groupe de classification

SECRETARY OF STATE. TERMINOLOGY DIRECTORATE. Les méthodes de travail du groupe de classification / The work methods used by the classification group

4 AGENDA

Monday, 2 April 1979

Welcoming address (Mr. Weis)

Opening address (Mr. Felber)

Morning Session (Chairman: Mr. Weis)

Presentation and discussion of the paper

Rondeau: Typologie des banques de terminologie

Afternoon Session (Chairman: Mr. Bachrach)

Presentation and discussion of the paper

Sundström: Transfer modalities of terminological data (presented by Mr. Algotsson)

Tuesday, 3 April 1979

Morning Session (Chairman: Mr. Landry)

Presentation and discussion of the paper

Brinkmann: Quality criteria for the exchange of terminological data

Special Session for Terminological Data Bank Managers (Chairman: Mr. Felber)

Discussion of problems concerning the management of terminological data banks

Special Session of representatives of organisationsplanning a terminological data bank (Chairman: Mr. Sager)

Afternoon Session (Chairman: Mr. Felber)

Presentation and discussion of reports.
Final recommendations

Bothe: The systematic classification of terminology / La classification systématique des stocks terminologiques

Dovbenko: Gosstandart activities in terminology and terminological data banks interaction

Goffin: The vocabulary of terminological data banks

Krommer-Benz: TermNet and INTERCONCEPT

5 WELCOME AND OPENING

5.1 WELCOME

Mr. Weis welcomed the participants on behalf of the Austrian Standards Institute and the Federal Economic Chamber. He then asked Mr. Felber to outline the objectives of the Conference.

5.2 OPENING

Mr. Felber expressed his pleasure that Mr. Löhner of Unesco and Mr. Dübell of the Austrian Ministry for Building and Technology had been able to attend the Conference. He thanked Unesco for the financial assistance for this Conference and the Ministry for Building and Technology for the generous financial support granted to Infoterm since its foundation.

Mr. Felber also thanked all those who had accepted the invitation to come to Vienna. He expressed his particular gratitude to those who had prepared papers or reports for discussion.

For this Conference a special group of experts, in particular representatives of terminological data banks in operation or in development had been invited. He stressed that there was no intention to create a Club of Vienna for terminology, restricted to a few terminologists and to exclude others. The group was to be kept small in order to permit discussion of certain important issues with a view to elaborate recommendations for further common action. For reasons of economy English was to be used as the sole conference language.

Owing to the great number of subject fields and their interdisciplinarity as well as the great number of languages to be covered close co-operation of all concerned was necessary.

Mr. Felber then stressed the necessity of co-operation with the suppliers of terminological data, i.e. the subject specialists and professionals in subject fields who are the most important partners of terminological data banks. Without their co-operation the quality and reliability of terminological data would be low. Great efforts would have to be made to ensure continuous co-operation of terminology preparing bodies and individual subject field terminologists.

This first conference had the purpose of discussing essential issues such as types of terminology data banks, data transfer, quality, classification etc. The topics to be dealt with were interrelated. In addition, a special session designed to further the exchange of views would be arranged for the managers of terminological data banks.

The outcome of these discussions should result in recommendations for the guidance of all those developing or improving terminological data banks, and in proposals for the solution of common problems. These recommendations might serve as a basis for UNISIS guidelines for the development of terminological data banks.

6 PRESENTATION OF PAPERS AND OF REPORTS WITH THE ENSUING DISCUSSIONS

6.1 EXPOSÉ D'INTRODUCTION

G. Rondeau
Université Laval,
Québec, Canada

TYPOLOGIE DES BANQUES DE TERMINOLOGIE

0. Introduction

Dans le document que j'ai soumis, je suggère sous forme schématique 6 points ou 6 aspects sous lesquels peuvent être considérées les banques de terminologie, dans l'optique d'une coopération internationale. Un septième point aurait pu être ajouté, celui du mode de traitement des données, mais je n'ai pas cru utile de l'inclure pour les raisons suivantes:

a) le mode de traitement des données, par lots ou autrement, est une question de régie interne qui exerce une faible influence sur la transmission des données d'une banque à l'autre et qui est plutôt liée aux objectifs propres d'une banque et à l'ampleur de la masse des données qu'elle doit traiter. Ce fait a été mis en lumière par les études des deux dernières années du G.T. 4 de l'ISO, qui se propose de recommander un format de transmission des données qui n'affecte pas la structure interne des fichiers mais sert plutôt à mettre en rapport les organes périphériques des diverses banques;

b) le but de TermNet n'est pas d'établir un ensemble de filiales compatibles entre elles et construites sur le même modèle, mais de mettre à la disposition de tous ses membres les résultats des travaux de chacun, ce qui devra nécessairement tenir compte de la diversité d'orientation et de fonctionnement de chacun des membres;

c) en supposant même qu'il faille un jour aborder l'aspect de la normalisation des structures internes de fichier cette question ne semble pas prioritaire dans l'immédiat - elle serait de toute façon irréalisable à court ou même à moyen terme -; je crois que nous devons plutôt, ayant admis la diversité des partenaires de TermNet, consacrer nos efforts de réflexion et d'imagination à la conception de moyens souples et efficaces de les relier entre eux.

0.2 Il y a également un 8e aspect que j'aurais pu signaler, soit celui des banques de terminologie dans leur degré de réalisation au sens étymologique, dans lequel on peut distinguer 3 types: a) les banques de terminologie existantes, b) les banques de terminologie en devenir, c'est-à-dire celles qui font l'objet d'un plan de développement et qui, à plus ou moins long terme, ont des chances de se matérialiser, et c) les banques de terminologie sur papier. Mais cet aspect sera, je n'en doute pas, étudié demain, au cours de la réunion des gestionnaires de banques de terminologie. C'est pourquoi je ne crois pas utile de l'aborder en détail maintenant.

1.0 La diversité

C'est là le problème fondamental, celui qui est la cause directe de la présente conférence, celui qui nous réunit ici ce matin. La solution de ce problème se trouve d'abord et avant tout dans une prise de conscience véritable et honnête du phénomène de la diversité, suivie d'une étude sérieuse de ses caractéristiques. Ce n'est qu'à la suite de cette double opération qu'il sera possible pour les gestionnaires des banques de terminologie, d'établir concrètement des modalités de coopération internationale. On comprendra dès lors la nécessité d'une discussion sur la typologie des banques de terminologie comme élément préalable aux discussions portant sur le contenu et les modes d'échanges.

Le postulat de la diversité peut s'exprimer sous forme des propositions suivantes:

I La mise en route d'une banque de terminologie exige une mise de fonds considérable qui fait suite à une décision d'ordre politique ou économique et doit, en conséquence, répondre de façon efficace à des besoins spécifiques.

Il découle de cette proposition deux corollaires:

1. Aucune banque de terminologie ne peut avoir comme objectif prioritaire la collaboration internationale en matière de terminologie (à moins que cette collaboration ne soit son objectif spécifique, ce qui n'est le cas d'aucune banque existante).

2. Toute banque de terminologie devra éventuellement, en vue de mieux répondre aux besoins spécifiques pour lesquels elle a été créée, faire appel aux échanges internationaux, étant donné la masse sans cesse croissante des termes à répertorier, classer et diffuser et étant donné l'impossibilité pratique pour une seule banque de traiter tous les termes.

II L'existence des banques de terminologie étant fondée sur la nécessité de répondre à des besoins spécifiques, il s'ensuit que chaque banque aura ses propres objectifs et, partant, ses propres critères de choix des données à traiter ainsi que ses propres méthodes de collecte, traitement et diffusion des données terminologiques.

Cette seconde proposition comporte à son tour deux corollaires:

1. Il y aura inévitablement des écarts considérables aussi bien

que des chevauchements dans le choix et le mode de traitement des données, pour différentes banques de terminologie.

2. Ces différences spécifiques ne sauraient, objectivement parlant, affecter en rien la réalité terminologique, qui doit demeurer identique à elle-même, qu'elle soit traitée au Canada, en Scandinavie, en France, en Allemagne, etc.

III Cette réalité terminologique est linguistiquement complexe et comporte les caractéristiques suivantes:

a) pour une langue donnée, à une notion correspond normalement un terme et un seul;

b) cette relation d'univocité entre un terme et une notion est propre à chaque langue et ne correspond pas nécessairement à une relation identique dans une autre langue pour une notion donnée.

IV Bien que la réalité terminologique soit, objectivement parlant, toujours identique à elle-même, chaque banque de terminologie privilégie ou met en valeur tel ou tel aspect de cette réalité en fonction des besoins particuliers de la clientèle à laquelle elle s'adresse.

Cette proposition comporte à son tour les corollaires suivants:

1. Les échanges entre banques pourraient être facilités si toutes les banques adoptaient le même sous-ensemble minimal d'éléments essentiels à traiter pour chaque type de données (c'est le cas pour les fichiers terminologiques qui se conforment aux accords de Baie St-Paul, 1972).

2. Aucune des banques de terminologie existantes n'utilise le même système de classement des données terminologiques; les systèmes de classement des données documentaires diffèrent également; un système uniformisé ne pourrait que faciliter grandement les échanges internationaux.

V Toute banque de terminologie assigne, implicitement ou non, deux types de valeurs relatives aux données qu'elle contient: une valeur relative interne se rapportant à une norme à l'intérieur d'une langue et une valeur relative comparée mesurant la qualité de l'équivalence notionnelle entre deux termes de langues différentes.

Deux corollaires découlent de cette proposition:

1. Les critères d'attribution d'une cote de valeur relative interne peuvent varier considérablement d'une langue à l'autre; en conséquence, leur harmonisation faciliterait grandement les échanges interbanques.

2. Toutes les banques de terminologie ne s'intéressent pas au même titre à l'attribution d'une cote de valeur relative comparée (voir par exemple, le cas de la BTQ).

Voilà cinq propositions que je considère comme fondamentales dans l'étude du phénomène de la diversité. Si l'on désire trouver des moyens de faciliter les échanges internationaux, il m'apparaît impérieux de classer ces aspects en deux catégories: d'une part ceux dont la diversité est irréductible et inévitable et dont il faudra nécessairement tenir compte dans l'aménagement d'un réseau d'échanges et d'autre part ceux qui pourraient faire l'objet d'une harmonisation internationale sans affecter la spécificité de chacun des membres du réseau. Mais avant d'aborder cette étude, regardons de façon un peu plus détaillée à l'aide du schéma qui vous a été distribué les points sous lesquels peuvent se regrouper les caractères propres des banques de terminologie. Nous pourrons, au cours de la discussion qui suivra, ajouter à ces points, en modifier la disposition, etc.

1. Objectifs
   - circulation des CST*
     - traduction
     - rédaction de textes
   - diffusion des terminologies
     - enseignement des Lsp*
     - implantation sociologique des terminologies
   - normalisation ou harmonisation des terminologies

---

* CST = Communications scientifiques et techniques
  Lsp = Langues de spécialité

2. Clientèle
- traducteurs
- terminologues et lexicographes terminologiques
- rédacteurs techniques
  - rédacteurs de CST
  - procès-verbalistes
  - rédacteurs de manuels, directives, etc.
- spécialistes de domaines scientifiques et techniques
- enseignants de Lsp
- grand public

3. Attitude linguistique
- normative (ne comprenant que des données normalisées)
- informative (comprenant des données non normalisées accompagnées de jugements de valeur)
- purement descriptive (comprenant des données de tous ordres, sans indications de valeurs)

4. Nature des données
- terminologiques
- documentaires
- traductionnelles (p. ex. quant au contexte d'emploi, quant à la correspondance notionnelle entre deux équivalents de langues différentes)
- grammaticales
  - morphologiques
  - syntaxiques
  - orthographiques
- encyclopédiques
- lexicales de langue commune
- visuelles (schémas, illustrations, etc.)

5. Organisation des données
   - fondée sur des documents
   - fondée sur des termes
   - quant aux langues
     - raccords bilingues
     - raccords multilingues

6. Modes de diffusion
   - en direct
     - terminal
     - téléscripteur
     - écran cathodique
     - par l'intermédiaire d'un terminologue (téléphone)
   - en différé
     - imprimante rapide
     - composition typographique automatique

2.0 Les solutions

En résumé, et pour conclure, car l'important, c'est de nos discussions qu'il sortira, si nous désirons mettre en place un réseau international d'échanges de données terminologiques et documentaires, nous devrons nous atteler aux tâches suivantes:

1. définir clairement les points sur lesquels la spécificité de chaque banque doit nécessairement être conservée et mettre en oeuvre des moyens de transmettre des données assorties d'une information concernant ces caractères spécifiques. On peut

mentionner par exemple sous ce chapitre le choix des données à traiter, les méthodes de traitement, la plus ou moins grande importance à apporter aux renseignements d'ordre traductionnel (et non seulement terminologique), etc.;

2. identifier les points qui peuvent, rapidement ou dans des délais raisonnables, de même que sans frais excessifs et sans porter atteinte à la spécificité de chaque banque, faire l'objet d'une harmonisation internationale, dans le but de soumettre aux membres du réseau des propositions concrètes;

3. établir des modalités de transmission de l'information selon les deux types mentionnés plus haut, soit en accès direct, soit par mode de transfert sur bandes magnétiques. Le premier type est évidemment lié à une question de disponibilité de lignes de transmission et de coût et son usage sera forcément limité à des cas spéciaux. Le second type pourra être mis en usage dès que l'ISO/CT 37/GT 4 aura mis au point un outil souple et facile d'emploi.

Ces trois étapes pourraient être, à mon avis, amorcées au cours de la présente conférence et confiées par la suite à des comités dynamiques travaillant selon un échéancier raisonnable.

4. La quatrième tâche concerne plus spécifiquement les gestionnaires des banques de terminologie. Il leur appartiendra, en effet, de mettre au point des mécanismes de compensation qui permettront à tous les organismes membres, grands ou petits, de se joindre au réseau, soit comme donneurs et receveurs,

soit comme receveurs seulement. Il va sans dire que ces mécanismes devront être équitables, en ce sens que le poids financier du réseau devra être réparti de façon proportionnelle sur tous ses membres. Une réunion est d'ailleurs prévue ultérieurement, au cours de laquelle ces questions pourront être abordées, ainsi que d'autres comme les questions de droit d'auteur, etc.

D'ici là, nous avons passablement de pain sur la planche. Je vous remercie de votre attention.

DISCUSSION

In the ensuing discussion the following points were made:

- phonetic transcription might be a desirable additional feature;
- banks should have an open-ended format to include additional categories;
- tables 4 and 5 of the appendix to the paper represented overlapping categories or criteria of differentiation;
- would there be any place for words in term banks?
- different user needs account for different content and structure of banks, i.e. translation, terminology work, documentation. What are the minimum data required for all three types?
- there already exists an agreement on minimum categories required for a bank. This agreement was reached in 1972 (Les données terminologiques. Actes du colloque international de terminologie, Baie Saint-Paul, 1er, 2 et 3 octobre 1972) among the banks existing at that time. Further agreements are desirable.
- the source language should be immaterial;
- a distinction was drawn but not elaborated - between terminology banks, terminological data banks and terminological banks (Goffin);
- the first of these should have contextual information;
- there are two modes of using data received from other banks:

  One is to use them side by side with the existing ones, the other is to incorporate the data into the bank; the second of these alternatives recommends itself for frequently used data;
- remote on-line access to banks is costly and often unreliable;
- it is important that users contribute to the data and quality control of holdings;

- a clear definition of user needs has to precede a definition of the categories of entries required;
- to the Third World access to terminological data is a problem of technical possibilities and know-how rather than a question of funds. Therefore it should be involved in all terminology activities in order to acquire new terminologies and terminological data as early as possible;
- agreement should be reached on three basic questions

  the form of the entry

  the coding of sources

  the subject classification used,

  this could be achieved by harmonization or even by standardization;
- formulas are desirable for re-formating data so that they can easily be incorporated into other systems.

Mr. Rondeau explained that his paper was a working document for discussion rather than an absolute statement. It was unlikely that existing banks, which respond to particular needs, would change their data structure; new banks could be planned to be more open to admit new categories of ordering and, therefore, new uses, especially via networks. User orientation was most important; he favoured separate holdings rather than incorporation of holding of one bank into another and also more remote access.

6.2 INVITED PAPER II

E. Sundström
SAFAD Fack
S-100 26 Stockholm, Sweden

TRANSFER MODALITIES OF TERMINOLOGICAL DATA

Contents

---

1 INTRODUCTION AND ACKNOWLEDGEMENTS

A prerequisite for better international co-operation in terminology is the free flow of terminological data between the various systems in current use. Several obstacles to-day impede that desired state of affairs: economical restrictions, language barriers, information gaps, differences in goals, technical incompatibilities etc.

Work aiming at the elimination of even the most trivial of these obstacles, the technical differences between the systems, must to a greater or lesser extent take each of the other ones into account. Apart from problems in data processing and telecommunication, questions of the equivalence and interrelation of data elements rank high among the technical issues involved in the interchange of terminological data. Here, arguments are raised in favour of an exacting attitude to the establishment of a content category scheme. Such a scheme would serve as a point of departure for future agreements on required data elements in international terminological interchange. The aim is to put forward a list of topics for further discussion rather than to suggest ready-made solutions.

It is appropriate to mention here that most of the views and ideas below originate from a joint Danish-Finnish-Norwegian-Swedish endeavour to establish communication between term banks, the so-called NORDTERM 2 project. The major standardizing achievement in the field, the MATER format, has not been explicitly tested in the project. However, it has been studied and elicited valuable suggestions. It is the intention of partners in the NORDTERM 2 project to submit an account of the Nordic experience in term bank communication at the Conference. Also, a comparison is to be prepared, with comments, between content category schemes used in different term bank systems. These forthcoming documents can be regarded as preliminaries for a joint Nordic proposal on required data elements in international terminological interchange.

2 A SURVEY OF THE PROBLEM

Interconnection of different terminology data banks is a frequently emphasized desideratum in current efforts to bring about efficient international co-operation between organizations active in the field. Interconnection can certainly be achieved in a number of ways, ranging from occasional exchange of machine-readable material to a coalescence of the various systems into one single computer network. However, regardless of the chosen level of interconnection, the first and foremost demand on the body administring the individual term bank is the establishment of a content category scheme for the fundamental entity in terminological work, the term record.

Such a scheme, typically, distinguishes between extra-linguistic and linguistic categories, mandatory and optional categories, source language and target language categories, morphological and semantical categories etc. It frequently recognizes hierarchical relations between categories and will in most cases introduce restrictions on repeatability, length, status etc. of the entries corresponding to the various categories.

This scheme, while presumably perfectly well suited to fulfil the various functions of the activity where it originated, should at the same time serve as a vehicle of communication with other activities which are likely to pursue quite different goals. Herein lies a source of internal conflicts in the scheme.

This situation calls for a thorough investigation on a basic level. Which, precisely, are the functions of the category scheme, in the internal processing and in terminological data interchange? Which are the universals of terminological work, and how are they reflected in the category scheme? In want of both the means and the capacity to realize such an investigation, the present inquiry will concentrate on more superficial aspects of the category scheme. But even so, it may be meaningful to account for different fundamental approaches to the problem. Here, a dichotomy is recognized between what may be labelled as an analytical and a synthetical approach.

2.1 The analytical approach

The conflict inherent in contrasting uses of the category scheme can be partly resolved by a distinction between different phases in terminological work, like e.g. the acquisition, processing, exchange and presentation of terminological data. Properties of the scheme which are necessary in the exchange phase are then singled out and treated separately. This attitude is equivalent to transfering at least some of the problems in terminology data interchange to the internal operations of the individual co-operating body. This is expedient, at times possibly even necessary. But it should never be allowed (and has not, so far as known, ever been used) to disguise the full scope of the task: it should be possible for each co-operating body, once a set of relevant data has been localized at some other body, to transfer that set into one's own system, to process it as if it were generated there and to make a presentation of it in any optional admixture with other data.

Anyhow, the problem is formidable enough, even if delimited to the drawing up of a widely applicable category scheme for the exchange phase in terminological work. Such an endeavour presumably proceeds from an analysis of an imagined, all-embracing term record all the way down to its most minute details. A comprehensive list of categories will be the result, at best forming an ordered set according to some consistent rules. So far, so good. But a difficult part of the task remains to be performed. Each co-operating body is required to match its own categories to that of the exchange scheme. This is certainly a rewarding work in many respects. Moot points can be scrutinized and remedied. Unnecessary complications, as well as over-simplifications, are likely to be discovered and eradicated.

But it should not be denied, either, that certain dangers are inherent in the analytical approach as delineated above. Much mental effort runs the risk to be wasted on the interpretation of intentions behind various categorizations, effort that might pay off better if directed towards a statement of the manifest and latent needs, leading to a deeper harmonization of actual terminological work by the different co-operating bodies.

After all, the analytical approach is equivalent to defining a set of labels: X, Y, Z,... The body administring the individual term bank seeks to establish consistency with its own set: x, y, z,... Ultimately, this is presumed to result in a series of equalities: x = X, y = Y, z = Z,... It may seem like an unwarranted insolence to observe here that equality is merely a special instance of difference. But that proposition is nevertheless singularly appropriate in this particular case. Every effort to deepen the analysis of the equality x = X will introduce a risk of bringing new aspects of x, and of X, to the surface and thus to disturb a probably hard-won equilibrium.

2.2 The synthetical approach

At least if superficially viewed, the antithetical approach to the problem would be to take into consideration primarily the influencing factors and the consequences of terminological data interchange. This attitude is here, for convenience, called the synthetical approach. For the moment it would be disregarded how, exactly, the interchange is brought about. Hints could thus be derived for a restating of the problem which in its turn might yield new insights into the part played by the category scheme in the transfer process. This is equivalent to letting observation pass via contemplation into combination (to cite a well-known formulation of the theorizing procedure. Theory is, after all, not restricted to scientific inquiry).

Perils of the synthetical approach are seen mainly in the direction of undue standardization of operations at the various co-operating terminological bodies. This might, it is felt, lead to a kind of officially adopted model of terminological work which by its very existence may be detrimental to a free and sound development. It is important to note that differences in attitudes, experiences and modes of operation are valuable ferments in international terminological co-operation and should be allowed to remain so. Interchange procedures should be designed with enough flexibility to let each body pursue its own goals.

Evidently, the analytical and the synthetical approach are complementary, rather than opposed to each other. The dichotomy has been treated here at some length because the synthetical approach is deemed to have been given too little attention in currently available proposals devoted to the subject. Also, the synthetical approach is recognized to be just as fundamental as the analytical approach and the title of this paper calls for a justification!

2.3 Influencing factors

A brief catalogue of influencing factors is presented below, with the purpose to give a rough conception of observations likely to play a part in the theorizing procedure referred to above. The factors have here been treated chiefly as impediments to the free flow of terminological data. A more optimistic note could easily have been struck, transforming the factors into incentives urging the development forward. But a more pessimistic outlook is felt to bring out the problems in full relief and thus, possibly, to stimulate further discussion of the issue.

2a) Economical restrictions are here taken not so much as a question of securing sufficient financial means for the implementation of new procedures. This is difficult enough, but should rather be viewed as a necessary moderating force than as one of the essential influencing factors. However, economy does play a more important part, inasmuch as there is always at least two partners involved in each single instance of data interchange and the costs incurred must be divided between them with some degree of fairness. In the typical case of information exchange to-day, most of the work is performed by the giving part. The receiving part will be required only to initiate the process by expressing wishes and, sometimes, to finalize it by paying the bill. An ideal interchange procedure should place all economical decisions on the receiving part and, if at all possible, not incur any costs by the giving part. Only the receiving part can assess the true value of the information retrieved and adjust actions accordingly. This attitude would seem to suggest a mode of interchange involving remote access in the search phase and data transfer by telecommunication followed by the necessary transformation process at the receiving end of the communication line. At least, this may be an adequate procedure in cases where piecemeal retrieval and transfer is the natural mode of operation.

2b) Information gaps, here taken to mean discrepancies in the understanding of identical situations between different human beings, is certainly worth every possible consideration at all levels in the communication of information. Here, to take but one example, reference can be made to the large gap in knowledge and aims between different categories of users. Certainly, most bodies administring term banks must develop their systems with a view to realizing decentralized input of data. This means that people of the most varied background must have at least a chance to make out how their data might fit into the category scheme.

2c) Differences in goals between co-operating terminological bodies have been referred to several times above. Of particular interest are issues in connection with the distinction between word-oriented and concept-oriented term records on the one hand, and between monolingual and multilingual term records on the other hand. It is in the very nature of terminological work proper to give precedence to the concept-oriented monolingual term record. However, other priorities must be recognized as legitimate in other contexts. Possibly, the need for interchange on a large scale is much more acutely felt in quarters dealing chiefly with the word-oriented bilingual term record. In such cases, the transfer procedure can probably be very straight-forward and perform well without a full-fledged content category scheme. The two groups of term banks thus distinguished may use partly different sets of procedures, depending on the relative size and number of term records to be interchanged. This should not be taken to mean, however, that the interconnection between the two groups should in any respect be less important than that within each one of the groups.

3 THE CONTENT CATEGORY SCHEME

With the need for investigations on a basic level duly emphasized, and with some of the important factors affecting these investigations treated in an albeit superficial manner, it might be excused if this short statement jumps to a few tentative conclusions.

3a) Theoretical aspects of the category scheme are seen to constitute a major difficulty in so far as the scheme cannot, indeed, be said to describe any obvious reality. On the contrary, its existence will to a large extent condition reality. This fact must always serve as an admonition to system-builders in the field. An allegedly flexible scheme which consists of an actually quite rigid system, whose flexibility rests in its faculty to allow for ramification ad infinitum, may ultimately prove to be detrimental. At least, it should be seriously inquired at an early stage in the development, whether a truly flexible category scheme could not be construed, giving maximum free scope to creativity in terminological work and at the same time preserving the possibility of unhampered data interchange. Inevitably, such a scheme (if it can be outlined with some credibility) will exhibit a compromise between rigour and freedom. It may be that a favourable combination of well-founded subdivison of the term record content on one hand, and open-minded standardization of the data entered into the categories on the other hand, will meet these requirements.

3b) Subdivision of the term record content is here taken to mean a plain delimiting procedure, without any bias towards some particular language or language family, without introducing any restrictions on the data entered into the categories and without any assessment of the relative value of different categories. The subdivision should be carried down to a level equivalent to content categories which are widely accepted as corresponding to required data elements in the acquisition, processing, interchange and presentation of terminological information. Thus is introduced a measure of convention in the scheme. Future shifts in information needs and habits will lead to corresponding modifications at precisely this level.

3c) Standardization of the data entered into the categories is here recognized as the main technique to evade excessive determination of details in the category scheme. It has always been a good rule in information processing to create a separate data field for each separate type of information. However, below the level recognized according to item 3b) as corresponding to required data elements, it might be advisable to dispense with this rule. It should be feasible, on this lower level, to standardize the form of the data entered into the various categories. Thus, the data will implicitly carry indications of its association with one or the other possible further subdivision of the category at issue. This technique transfers much of the complications from the category scheme to the procedure which decodes the data. The advantage is that the vast majority of terminologically active bodies dealing with comparatively straightforward data never needs to bother with more complex data structures or procedures. Still, their data can be taken care of by co-operating bodies engaged in such complex processing, provided that the data comply with certain widely diffused rules of writing.

—

As an illustration to the line of thought pursued in this section, the Appendix presents an outline of required data elements in terminological interchange.

4 PROPOSITIONS ON CONTENT CATEGORIES

At this juncture, a medley of topics present themselves for contemplation. The topics below do not explicitly refer to the general picture of transfer modalities advanced in this paper, and should be possible to discuss by all interested parties without any commitments. The topics are here presented in the form of propositions, framed with a view to inviting comments and certainly not with pretensions to be exhaustive or conclusive.

4a) The categories are created in terminological practice.
No transcendent sources exist, from where the content categories are derived. They should rather be regarded as embodying usage and, as such, being susceptible to change according to shifts in information needs and habits in society. This simple observation is here intended as a support to the following two propositions.

4b) Identity in the categories is always relative.
Equivalence has to be forced upon categories belonging to different systems, just as equivalence is forced upon words belonging to different languages. And, to carry the comparison one step further, each category derives its identity from the existence and relative position of other categories in the same system.

4c) The categories are manifestations of intentions.
To be sure, the categories do act as a classification means. This aspect is treated in the following two propositions. The primary part to be assigned to the categories in terminological work is, however, as expressions of intentions in relation to the aims and operations of the body administring the term bank in question. The terminological data should be regarded more as a vehicle of creativity and competence than as something static.

4d) The categories should form a perfect hierarchy.
By this is meant that each category should be defined as part of one and only one other category on the next higher level. This requirement is easy to fulfil in a trivial manner (by appointing the entire term record to be level 0 and all other categories to constitute level 1). It should be observed, however, that this proposition is conditioned by the following one.

4e) Each category should be substitutable by the next broader one.
A measure of freedom is thus given both to those co-operating bodies who do not need to work out their categories below a certain level, and to those who for some reason choose to specify the scheme in all its details. To be sure, this principle might lead to imperfect matching at times. But the imperfection is at least under control.

4f) Data entered in the categories are pure terminological data.
Introducing separators or similar information in the categories is either unnecessary (whenever this can be performed by the system), or equivalent to establishing new implicit categories (which may be convenient and should not be regarded as in conflict with the proposition). It is true, however, that provision must be made for information of a meta-terminological kind, like for instance messages from one co-operating body to another.

4g) No restrictions apply to length, character repertoire etc.
It is obviously convenient, in internal processing, to allow for certain restrictions on data entered into specific categories. However, no such restrictions should be prescribed in a scheme for interchange purposes.

5 TOWARD A FREE INTERCHANGE OF TERMINOLOGICAL DATA

To yield useful results, discussion on the structure and application of a content category scheme for terminological data interchange should take into consideration both fundamental issues, like the ones touched upon above, and practical experiences. As for the latter aspect, a procedure is here advocated for systematic feedback from actual implementation of interchange on a national, regional or international level. This procedure calls for concerted action according to a coherent strategy. Here, the content category scheme has been recognized as the key issue in any attempt at terminological data interchange. But it is probably wise to broaden the field of vision, to include other and more practical problems likely to influence the forming of a widely accepted platform for further development in the field. Thus, presuming that a content category scheme has been decided upon, the following observations may be relevant.

5a) The computer format and the category scheme are independent.
The details of the information processing format, including requirements on record labels etc., depend to a great extent on physical circumstances such as computer system, magnetic media etc. Interchange implies agreements on these details and how they should vary according to the circumstances. Care should be taken to keep the pertinent decisions free from undue ties to the content category scheme. The point is that development in information processing and in terminological work follow different paths. Both the processing format and the category scheme are subject to change. It is essential that complications arising from unnecessary intermingling of the two fields of endeavour do not hamper the overall development.

5b) The category scheme should be applied to all terminological work.
The very concept of content category scheme calls for an ambition on behalf of all bodies administring term bank systems to apply the agreed scheme to each separate phase and each single task in the collection, storage, processing and dissemination of terminological information. The scheme should be amended, not abandonned, whenever it proves to be inadequate in some respect. Now and in a foreseeable future, the free flow of terminological data within an individual body is far more important in economic terms than the free interchange between co-operating bodies. Only at a stage characterized by great interdependence and division of labour will this state of affairs undergo a significant change. It is profitable to the internal operations, and advantageous to future co-operation, if each body administring a term bank system takes these circumstances into due consideration.

5c) An existing category scheme will bring other issues to the fore.
Under this non-committal heading is subsumed a host of problems too far-reaching to warrant other than a cursory mentioning in this context: questions of copyright and data safety, fees and remuneration, combination of information retrieval of terminological data and that of a more general kind etc.

Appendix
REQUIRED DATA ELEMENTS IN TERMINOLOGICAL INTERCHANGE

A Scope and application

This specification of data elements in terminological interchange rests on an imagined, flexible content category scheme which is not borne out in detail here. The specification is intended only to illustrate arguments and conclusions in the paper "Transfer Modalities of Terminological Data", particularly those in Section 3. This document is not identical with the prospect joint Nordic proposal in the field, referred to in Section 1 of that paper. The term records, subject to rendering by means of the data elements, are presumed to be monolingual, concept-oriented records.

B Data elements

Each data element is here assigned a number, a name and a definition. When applicable, notes are given on form and content of the data entered. The list of data elements is not intended to be exhaustive, for reasons stated under item A above.

1 origin
association of the term record to a source, a project work, an organization etc.
In the case of excerpting from e.g. a standard reference work, the abbreviated form of the title will be sufficient. An individual term bank system will typically use some formalized notation of the project work at issue, possibly combined with a subject code. A specialized organization with terminological activity restricted to its own subject field will probably give its acronym only.

2 reference number
serial or otherwise regularly varying number identifying the term record within the source, project work etc.
Items 1 and 2 in combination will unambigously determine the term record.

3 date
the date the term record was entered, established, modified etc.
The date shall be written as specified in ISO 2014, e.g. 1979-02-13. The exact meaning of the date entry may be signalled by accompanying codes.

.
.
.

11 main entry
the linguistic expression of the concept which the term record is intended to comprise
The main entry should be given in its basic form.

12 morphological information
description of the morphological properties of the main entry
The form and relative order of the data entered in this element should provide sufficient clues for an exhaustive decoding conforming to the needs of any co-operating body.

.
.
.

21 <u>definition</u>
a succinct description of the meaning of the main entry

.
.
.

91 <u>note</u>
meta-terminological information directed to members of the project group responsible for the term record, readers of the forthcoming glossary etc.
Text entered in this element will be interpreted according to the circumstance in which it is reproduced.

.
.
.

D Comments

The few examples of data elements given above should suffice to illustrate the basic tenet of the paper. Subdivision of the term record by means of a content category scheme, resulting in a selection of data elements, should not be carried beyond a level corresponding to the needs of a majority of interested parties. Further details can, when necessary, be implemented by means of a sufficiently standardized form of the data entered in the elements. This will imply agreements on the relevant issues between bodies engaged in terminological work of a more diversified character.

DISCUSSION

The chairman, Mr. Bachrach, had asked Mr. Algotsson in the absence of Mr. Sundström to present his paper. The following points emerged in the discussion:

- there is a danger of unreliable terminology being produced by the new so-called 'pocket dictionaries';
- it should be a principle that one category of a term record should contain only one type of information;
- the MATER format is not yet agreed upon and does not, therefore, represent a recommended format; it is only intended as an exchange format but may offer guidelines on possible categories for new banks; the discussion on this ISO Draft Proposal is far from complete and proposals exist for a reduced number of categories; it may be desirable to wait and let more experience be gained in this field before attempting to reach international agreement on categories;
- the existing MATER proposals are flexible in as much as they permit repeating categories;
- it is relatively easy to achieve agreement on the categories required to record standardized terminology;
- it may be desirable to limit terminological data collection to standardized terminology, thesauri, special vocabularies and multilingual vocabularies.

Mr. Algotsson replied that in his opinion it was diffcult to assign specific categories to rather deversified types of information.

Mr. Felber expressed the opinion that it might be desirable to approach international and national organizations working in terminology with a request to collaborate in the collection of terminological data. This might eventually lead these organizations to becoming users of banks.

Mr. Bachrach closed the session with the observation that publicity about the importance of terminology was highly desirable.

6.3 INVITED PAPER III

K.H. Brinkmann
Siemens AG,
Munich, FRG

QUALITY CRITERIA FOR THE EXCHANGE OF TERMINOLOGICAL DATA

**Mr. Chairman,
Ladies and Gentlemen:**

When I came to Vienna, I had - what I thought - a clear idea of what quality criteria as referred to the exchange of terminological data should cover.

After yesterday's discussions I am somewhat at a loss as to whether what I am going to tell you will be what you might expect to hear. Because I will restrict myself to stating what should be included in the exchange of terminological data to make the exchange worthwhile - to give it a sound basis. I will definitely not talk about the quality of the exchanged material itself - because determining the quality of the terminological data is a very subjective matter and depends on the individual points of view of the partners engaged in the exchange. So what I will do is talk again about types of information or information categories.

The paper I am going to read to you reflects the results of work done by a study group formed by members of a number of institutions and private companies cooperating in the field of terminology on the basis of a terminological data bank system viz. the Translation Department of the Dutch Ministry of Foreign Affairs, the Institute for the Training of Translators and Interpreters of the University of Saarbrücken, as well as the Translation Departments of N.V. Philips Gloeilampenfabrieken in Eindhoven, and Siemens AG in Munich.

As science and technology continue to advance at an ever quickening pace, with developments tending to become more and more abstract and less easy to correlate, mounting obstacles are being placed in the way of both interdisciplinary communication within one linguistic area and intradisciplinary communication between different areas, across language barriers. The transfer of knowledge from one linguistic areas to another is further hampered by the fact that we are not moving in the direction of having a small number of languages, or even a single "lingua franca", for the interchange of information. Instead, the growing national awareness being witnessed in many countries is having the effect of constantly increasing the number of languages used in the technical forum.

As a result of these developments, sound and thorough work in the terminology field is becoming increasingly more important. This can clearly be seen from the growing number of centers and institutions engaged in coining, collecting, standarizing and systematizing terminology. It is obvious that none of these organizations can encompass the entire terminology of every single field of specialized activity in all languages, even on a national let alone international scale. This fact has naturally given rise to specialization based on prevailing interests, which in turn means that terminology centers are unable to accommodate anyone requiring terminological information from fields outside their own speciality. On the other hand, a terminology center may find itself having to deal more and more with a particluar field or language which it has had nothing to do with before but in which other centers are experienced. In this case, the other centers would have a stock of useful information on the subject or in the language concerned.

For this reason it is desirable for different centers to be able to exchange terminologies. However, for this to be possible there are certain quality criteria which have to be fulfilled by all those interested in participating.

A point to be taken into account here is the fact that collections of terminology, or "terminology banks" as we shall call them without regard to their technical realization, can serve different purposes. For instance they can be oriented toward standardization, but they can also serve as a basis for the transfer of knowledge across language barriers, i.e. be oriented toward translation. At present the translation-oriented data banks are by far the largest, and will probably remain so for some time to come. Their purpose also makes them prime candidates for an exchange of data.

We shall refer to the exchangeable data as "terminological units" (TU) and consider what information must be contained in them as a minimum requirement for ensuring an acceptable level of quality.

The following information must be regarded as indispensable:

1. A code which clearly identifies the TU so that it can be altered, supplemented, reduced or erased if necessary even after the exchange has taken place.

2. A quality rating, i.e. an indication as to whether the information is fully reliable, reliable in part, provisional or tentative.

3. Date on which the TU was produced or last updated.

4. Person, body or institution which created the TU and who or which should therefore be referred to in cases of uncertainty.

5. The special field to which the TU has valid application. In this connection it should be noted that a uniform system of classification is unlikely to emerge in the foreseeable future. It seems that the majority of people working in the terminology field consider the UDC to be inadequate, too cumbersome and outmoded. All large terminology banks have their own classification systems which are chiefly aimed at meeting their retrieval requirements, so the recipient of terminological data must be informed of the system employed by the supplier. Anyway, this is a topic dealt with in Mr. Bothe's Report.

6. Term (single- or multi-word term, phraseological expression).

7. Source of the term.

Items 6 and 7 apply to each of the languages contained in the TU, of course.

Beyond these minimum requirements there is the following additional information which is also important and therefore desirable:

8. Definitions of the concepts to which the terms are to be allocated.

9. Sample contexts illustrating how the terms are used in the languages concerned.

10. A means of indicating which was the source term when the TU was compiled.

11. Is the TU in its present form limited in its validity (only valid in connection with a specific device, system etc. in the case of a term used in technology, for example)?

12. Are terms subject to regional limitations in their use (USA, GB, CH, A, D, etc.)?

13. Grammatical information on the designations, e.g. part-of-speech labels. Information on pronunciation.

14. Language level indicators (laboratory, factory, advertising, jargon etc.).

15. Recommendations regarding use (preferred, allowed, deprecated etc.).

16. In the case of multi-language TUs, an indication that a specific term in word lists, glossaries etc. derived from the terminology bank cannot be a source-language search word.

17. Relevant bibliographical or documentary information relating to the TU.

Not to be ignored are a few fundamental quality criteria which, for historical or other reasons, have not yet been met by all operational terminology banks. These include:

1. Correct orthography, i.e., with respect to those languages with alphabets derived from the Latin, use of capital and small letters, accents and all other diacritics. The reason why this is important is that the terminology banks are in most cases also used by people whose linguistic competence is not perfect.

2. Unambiguous coding and transliteration systems for all non-Latin alphabets, designed to be readily managed by those who receive the information.

3. Variable length of the TU and of all the information contained in it, a requirement which can by all means be fulfilled in a number of different ways in the individual terminology banks.

Everyone knows that much time has been spent on considering how to fulfill this list of quality criteria in practice and how to realize the exchange of data from the technical viewpoint. In this connection we may recall the efforts to obtain a standardized magnetic tape format for the exchange of terminology (MATER) and also the exchange criteria which have been established within terminology cooperative associations. To go beyond the topic of this paper it may be mentioned that relatively simple rules can be established for the exchange of terminology via data carriers, but that it will almost certainly be a very different matter when it comes to remote data transmission or even interactive mode operation. It would be a good thing if standards could be developed that are acceptable to all those involved.

In closing I should like to comment on a suggestion made yesterday by Ms. Gorm-Hansen of Copenhagen. I am much in favor of that suggestion, namely that Infoterm should get from all operative data banks a detailed and fully explained description of the individual information categories they use, maybe with an outline of further categories that they might consider useful in the future. This would give us an idea of what would be required in actual practice and save us from unnessessary ballast in our reflections on the requirements to be met in a worldwide exchange of terminological data.

DISCUSSION

The chairman, Mr. Landry, opened the discussion in which the following items were raised:

- with regard to data exchange the following distinctions have to be made: exchange of complete vs. partial records, the exchange of documentation vs. the exchange of term records, descriptive vs. prescriptive data, the internal vs. the external format of data, i.e. the exchange format;
- with regard to term records:

  the date of the incorporation of a term into a bank is important to identify new accessions; only terms originating from standards are of uniform quality; the source of a term is a useful guide as to its quality; the function of a bank determines its quality requirements, i.e. banks intended to assist translation have to accept provisional terms and translation equivalent coinages which will not be acceptable to a bank of standardized terminology;
- with regard to translation equivalents:

  translation equivalents are not necessarily full conceptual equivalents; ISO 1951 provides symbols to indicate degrees of concept correspondences, equivalences can be tested by comparing definitions, pictorial representation can assist in identifying the degree of equivalence; there is a danger of purist attitudes in deciding on equivalents;
- with regard to categorization:

  categories vary over time, categories are necessary for multiple use of data, to speed up search, for close identification of terms; there was no full agreement on the basic category requirements among the various papers presented at the conference; there was a need to study the essential categories required for all data banks both for the purpose of facilitating exchange of data and to give guidance to new banks. Unesco would welcome guidelines for new banks especially with respect to data elements to be included,

ISO discussions on an exchange format should be brought to a positive conclusion as soon as possible; other ISO recommendations were not being observed as they should be (Zdorovov);

- there is a distinction to be made between retrieval criteria and output information; the compatibility of retrieval criteria should be aimed at for the greater benefit of users;
- the statement listing the intended categories for the UN data bank was welcomed.

The chairman summed up the discussion by stating that the first priority was to obtain full information about the data stored in the various banks.

6.4.1 SPECIAL SESSIONS

The second session (morning of April 3) was divided into two working groups:

(1) the managers of existing banks. Chaired by Mr. Felber, it was devoted to practical problems as well as policies of terminology problems in general

and

(2) the members of the conference who are planning to establish a data bank.

The discussion of the latter working group, chaired by Mr. Sager, was concerned with:
- the need of clear definitions of user requirements;
- the possibilities provided by new technology in the form of word processing systems and micro processors for data storage;
- the important contributions which can be made by the users themselves;
- the role of banks in the teaching of special languages;
- the contribution banks can make towards the harmonization of language in large institutions.

The general conclusion was that much could be learnt from existing banks and that, therefore, guidelines based on existing experience would be of great assistance.

6.4.2 SPECIAL SESSION FOR THE MANAGERS OF TERMINOLOGICAL DATA BANKS

The special session of TDB managers was held to discuss problems connected with the management of TDB, such as:

(1) problems concerning data transfer:
indication of sources, copyright, restricted vs. unrestricted data transfer, quality criteria, nature of interrelations between TDB, e.a.

(2) problems concerning the cost of data management: structure, standardization, formats and classification, data acquisition, data marketing, financial modalities, e.a.,

(3) problems concerning the qualification of personnel: training, exchange of experts.

The discussion clearly revealed the need for better communication and for a code of good practice between terminological data banks.

6.5.1 INVITED REPORT

H. Felber
Infoterm
Vienna, Austria

REPORT ON TERMNET

1 Introduction

In autumn 1971 Infoterm was established in Vienna with the assistance of Unesco within the framework of UNISIST (now General Information Programme of Unesco); it was affiliated to the Austrian Standards Institute [1].

At the first Infoterm Symposium on "International co-operation in terminology" in April 1975 networking of terminological activities was discussed [2]. Infoterm was charged with the elaboration of a study on the development of a network for terminology information and documentation. This study was prepared by Infoterm in 1977 [3]. It was discussed, modified and approved by a group of experts convened by Unesco in Vienna in October 1977 [4].

In 1978 an inquiry about the participation in TermNet was made by Infoterm among organizations engaged in terminology all over the world. The result of the analysis of the questionnaires was distributed to those answering the questionnaire [5].

The result of the analysis of the questionnaires and a draft plan for data management within TermNet will be discussed at the next Infoterm Advisory Board Meeting.

At a future meeting,convened by Infoterm for all those who expressed the wish to join TermNet, details of TermNet such as membership, structure, exchange of data, formats, etc. will be discussed.

2 TermNet Programmes

TermNet comprises three programmes:

Programme 1: Developing the scientific basis for terminology (General theory and principles of terminology)

Programme 2: Establishing closer co-operation in preparing terminologies

Programme 3: Establishing closer co-operation in collecting, recording, processing, and disseminating terminological data and information

TermNet Programme 1 aims at concentrating the forces for terminology research and dissemination of the General theory of terminology. It also includes training, and the preparation of teaching material.

Some AILA-Infoterm initiatives have already been taken.

TermNet Programme 2 aims at the organization of continuous elaboration of terminologies in all fields by learned, professional and specialized organizations and experts. This is a very difficult task indeed. If these organizations will not become the main suppliers for terminological data banks in the near future term-banks will never be able to fulfill tasks assigned to them.

TermNet Programme 3 concerns the organization of co-operation in collecting, evaluating, recording, processing, storing, exchanging and disseminating terminological information. The terminological information comprises the following types of data:

(1) terminological data (such as terms, synonyms, definitions, context, foreign language equivalents, sources, code of quality, etc.)
(2) bibliographical data on terminological literature
(3) data on organizations engaged in terminology, terminology experts, terminology projects, etc.

Some terminological data banks have already established a co-operative scheme with respect to terminological data.

However, a decision will have to be taken, whether general co-operation within the framework of the TermNet Programme 3 (for instance as sub network) is desired and in which form.

On the bibliographical level the elaboration of International Bibliographies has been started. The following bibliographies are being prepared and kept up to date:

BT 1 International Bibliography of terminological papers and books. A pilot project is being undertaken by Infoterm, AILA (University of Laval, Quebec, Canada), Office de la langue française (Canada) and others to store bibliographic data on magnetic tape and publish them as soon as possible.

BT 2 International Bibliography of standardized vocabularies. These data are stored on magnetic tape in Infoterm. The bibliography was published 1979 as Infoterm Series 2 [6], it is being kept up to date.

BT 3 International bibliography of multilingual specialized vocabularies. It is prepared by subject groups. Two pilot projects are under way with the Economic Commission for Europe of UN (for building and construction) and with Interconcept (Unesco) (for the social sciences).

Infoterm is particularly active with reference to data on organizations engaged in terminology, on terminology experts, terminology projects, etc. In 1977 Infoterm published the "World Guide to Terminological Activities" as Infoterm Series 4.

In the near future it is intended to computerize this "World Guide" and to keep it up to date, if funds will be made available.

REFERENCES

[1] Unesco. Report on the meeting for the evaluation of the International Information Centre for Terminology (Infoterm). UNISIST Steering Committee, 5th meeting. Paris: Unesco, August 1976, 45 p. SC/UNISIST/ST.COM./BUR.5/3.

[2] INFOTERM. First Infoterm Symposium. International Co-operation in Terminology / Premier Symposium d'Infoterm. Coopération internationale en terminologie. München: Verlag Dokumentation, 1976, 332 p., A5. Infoterm Series 3.

[3] FELBER, H. Study on the development of a network for terminology information and documentation. Vienna, 1978, 45 p.

[4] Unesco. Report on the meeting for the evaluation of the Study on the development of a network for terminology information and documentation, prepared by Infoterm. Vienna: Infoterm, oct. 1977, 12 p.

[5] KROMMER-BENZ, M. Co-operation in TermNet. Analysis of Questionnaires. Vienna, 1978, 21 p.

[6] FELBER, H; KROMMER-BENZ, M; MANU, A. International Bibliography of Standardized Vocabularies / Bibliographie internationale de vocabulaires normalisés / Internationale Bibliographie der Normwörterbücher. München: Verlag Dokumentation, 1979, 540 p., A4. Infoterm Series 2.

[7] KROMMER-BENZ, M. World Guide to Terminological Activities / Guide mondial des activités terminologiques. München: Verlag Dokumentation, 1977, 311 p., A5. Infoterm Series 4.

6.5.2 INVITED REPORT II

A.A.P. Bothe,
Dutch Ministry of Foreign Affaires
The Hague, Netherlands

LA CLASSIFICATION SYSTEMATIQUE DES STOCKS TERMINOLOGIQUES

Quel est le groupe cible? Les traducteurs et les terminologues appartenant à des services de traduction possédant des stocks terminologiques d'une certaine ampleur. Les utilisateurs souhaitent manifestement une classification systématique qui faciliterait à la fois l'accès à ces stocks ainsi que leur contrôle et leur mise à jour.

La situation se présente aujourd'hui comme suit:

A l'origine, la constitution de stocks terminologiques s'est faite ordinairement sans classification systématique. On n'en reconnaissait pas encore la nécessité, et le recueil d'un nombre maximal de données terminologiques constituait l'objectif premier Le seul système appliqué était le plus évident, l'alphabet.

L'expérience montre toutefois que lorsqu'un stock emmagasiné (sur ordinateur) compte quelques dizaines de milliers de termes, l'alphabet comme seul moyen de sélection ne suffit plus. L'offre d'information sur l'écran ou sur la microfiche devient confuse, la recherche s'allonge, la distinction entre les nuances de contexte ou des unités terminologiques à première vue identiques (par ex. entre homonymes) sont moins aisées à discerner.

Bref, l'absence de classification systématique entrave la gestion efficace d'un stock sur les plans de l'actualisation, de la fiabilité et de l'extension orientée.

Quels sont les avantages d'une classification systématique de la terminologie?

Outre la possibilité d'une structuration détaillée, ou non, des stocks, possibilité qu'offre d'ailleurs tout système de classification, on pourrait citer l'intégration de terminologie dans les langues de travail cible, le traitement et l'actualisation systématique de stocks partiels et donc l'amélioration de la qualité de l'ensemble des stocks. Une classification pourrait également permettre une collaboration spécifique et une division du travail entre services de traduction et de terminologie par l'échange de stocks, partiels ou non. On pourrait ainsi encourager l'utilisation d'une terminologie uniforme, éviter les chevauchements et réduire les coûts. Enfin, il est évident que l'indication du secteur d'application d'une unité terminologique constitue une donnée utile pour l'utilisateur consultant un stock par voie alphabétique.

Une classification systématique doit être utile pour l'utilisateur, c'est-à-dire le terminologue et le traducteur. Le système doit donc être accessible sans instructions complexes et sans formation spécifique.

Cette simple constatation amène déjà à écarter les systèmes existants, notamment les systèmes documentaires. En effet, l'utilisateur ne peut s'en servir sans formation spécifique, la recherche est longue, la subjectivité dans le classement des données effectué par les spécialistes du système de classification cloisi entraîne inévitablement des pertes d'information.

En outre, il existe des différences fondamentales entre le classement de documents et le classement d'une unité terminologique. Dans la documentation, les informations résumées ou connexes sont classées par sujet ou en fonction de certains dénominateurs. Un document traitant d'un sujet particulier peut comprendre des notions terminologiques se situant dans des domaines fort différents, par exemple des notions mathématiques peuvent se retrouver dans des textes traitant d'économie, de biologie, de circulation, etc. Ces notions ne sont pas automatiquement inhérentes au sujet du texte, qui, par contre, constitue lui-même une unité immuable.

Les notions ne sont donc pas limitées à un seul sujet. Elles sont inhérentes au domaine de connaissances dont elles procèdent. Les notions peuvent changer de signification, en fonction de l'évolution du domaine dont elles relèvent. En théorie, la classification des unités terminologiques pourrait se faire selon les mêmes principes hiérarchiques utilisés dans la classification documentaire. Mais c'est irréalisable en pratique parce qu'une même notion peut se retrouver dans les domaines les plus différents et que le nombre de descripteurs serait alors très vaste. J'en conclus que l'ordonnance des sujets et l'ordonnance des notions ne peut se faire suivant les mêmes principes.

Divers services de terminologie en Europe et hors du Continent utilisent des classifications matières. Pour autant que je sache, ces systèmes n'ont aucun lien entre eux, ont été mis au point au sein même des services, et bien souvent une fois que les stocks avaient déjà atteint une certaine ampleur. Leur objectif premier était, tout naturellement, les besoins du service. Je ne connais pas bien leur historique ni la méthode suivie: s'agit-il de systèmes documentaires adaptés ou de systèmes nouveaux, concus en fonction de la structuration des stocks terminologiques? S'agit-il de l'adaptation ou de l'élaboration de systèmes par les utilisateurs, ou par les spécialistes de la classification, ou par les deux? A-t-on tenu compte de l'effet que pourrait avoir le système sur la rétroaction par le traducteur?

J'ignore également si l'on a déjà cherché à échanger entre services des stocks, partiels ou non, couvrant un même domaine pour les emmagasiner et les rendre accessibles suivant leur propre classification.

Je serais heureux que l'on procède à une étude approfondie de ces questions.

Pour autant que je sache, on ne s'est jamais efforcé d'élaborer un système de classification terminologique qui ne soit pas conçu en premier lieu en fonction des besoins de l'organisation mais qui ait un caractère "universel", c'est-à-dire un système qui pourrait être utilisé par tous les services de terminologie intéressés et qui permettrait l'échange de stocks sans perte sensible d'information. Peut-être l'absence d'un tel système s'explique-t-il de la façon suivante.

Il convient de relever en premier lieu qu'INFOTERM ne fonctionne que depuis relativement peu de temps comme banque de compensation, comme clearing-house, des données terminologiques. Auparavant, ce que l'on savait des activités terminologiques déployées ici ou là dépendait généralement du seul hasard.

Il en découle que bien souvent les domaines dans lesquels opère essentiellement un service ont probablement constitué le noyau autour duquel s'élaborait son système de classification, et que sans doute aussi les principes de classification documentaire déjà utilisés dans ce service ont exercé une influence directe ou indirecte.

On peut se demander s'il est souhaitable et possible de promouvoir l'élaboration d'un système de classification universel et spécifique pour la terminologie et mis au point jusque dans les détails. En théorie la réponse devrait être affirmative. Mais elle doit peut-être être négative sur le plan pratique. Je pense en particulier qu'il est impossible de faire élaborer un tel système de façon scientifique par les utilisateurs. Mais le faire élaborer par des spécialistes de la classification sera fort coûteux et exigera notamment une analyse approfondie des besoins des utilisateurs. De plus, l'élaboration d'un système universel demande une connaissance approfondie de la nature et de l'ampleur des stocks existants et devant être traités.

Il ne suffit pas de savoir que le service A possède 150.000 termes en matière de navigation spatiale et 80.000 en matière de théologie.

Pour quelle langue un tel système devrait-il être élaboré, ou pour quelle combinaison de langues? Les stocks existants doivent-ils être convertis pour pouvoir être intégrés dans de nouveaux systèmes?

On le voit, tout ceci implique des années de travail en commission de nombreux spécialistes de la classification, de traducteurs, de terminologues, de spécialistes de l'informatique.

Pendant ce temps, les stocks continueront à grandir.

Quelles perspectives offre 1984?

Est-ce qu'à l'avenir l'utilisateur devra chercher des données dans des stocks comprenant des centaines de milliers d'unités et avec pour seul outil de sélection l'alphabet? Ou bien devra-t-il faire appel à la fois à un spécialiste d'un domaine déterminé et à un spécialiste de la classification pour avoir accès aux glossaires dont il a besoin?

Je l'ai déjà dit, je n'apporte pas de solution toute faite. Il me semble toutefois que la collaboration d'un certain nombre de services de terminologie se complétant ou travaillant dans les mêmes domaines, aussi bien en largeur qu'en profondeur d'ailleurs, et qui chercheraient de concert une solution pragmatique pour la classification de leurs stocks, permettrait d'arriver à des résultats valables. Il convient, avant d'aborder la classification systématique d'un stock terminologique, quel qu'il soit, de définir nettement les exigences qualitatives et techniques auxquelles doit satisfaire une unité terminologique lors de son introduction dans le stock. En effet, c'est le terme et tout son environnement de données qui détermine l'attribution du code. M. Brinkman s'est longuement attaché à ce problème dans son exposé d'hier.

Quelles règles conviendrait-il de suivre lors de la classification de stocks terminologiques dans le cadre de coopération entre divers services? Il est improbable qu'il existe un service de terminologie dont les activités se répartissent uniformément entre tous les domaines de la connaissance humaine; il est beaucoup plus vraisemblable qu'il y a ordinairement un nombre limité de secteurs clé: il faudrait donc songer à une répartition du travail entre les organisations qui coopèrent, après avoir procédé à une classification sommaire en catégories des domaines intéressants du point de vue terminologique. Chaque partenaire est alors responsable de la structuration et de la gestion d'une ou plusieurs catégories. Il convient de s'entendre sur l'ampleur maximale de chaque catégorie, sur le nombre de mots clé pouvant être repris dans chaque catégorie et sur les principes de répartition et de division. Il convient de donner la préférence à la méthode inductive plutôt qu'à la méthode déductive - certainement si les stocks sont encore limités - lors du classement dans une catégorie. Il est vrai que la méthode déductive est plus scientifique, mais elle impose une préparation trop longue vu les principes théoriques qui doivent être établis au préalable, principes qui sont malaisés à adapter à l'évolution et qui ne tiennent pas compte de la dynamique propre aux notions à classer.

Il faut donc, pour éviter des pertes d'information et des informations trompeuses, attribuer un code en fonction du contexte signifiant dans lequel le terme a été trouvé. Cela veut dire, en principe, un seul code par unité terminologique, donc à la fois pour la langue de départ et la langue d'arrivée, au moment de l'attribution. C'est donc le contexte de la langue de départ qui détermine le codage de l'unité terminologique.

Ce principe présente divers avantages:

- c'est le contexte signifiant qui détermine l'attribution du code, cela facilite le choix du mot clé. L'équivalent dans une autre langue arrêté en fonction des données du texte de départ se classe sous le même mot clé;
- l'attribution d'un seul code par unité terminologique prévient que le codage soit déterminé par la culture générale et les connaissances du codeur et que des distorsions puissent découler de significations divergentes des termes de départ et d'arrivée en dehors du contexte premier;
- la décision du codeur peut être facilitée par la répartition des mots clé en fonction d'un certain nombre d'aspects, par exemple la terminologie spécifique, les fonctions, les organisations, la législation pertinente, compte tenu des besoins des services. Les divers aspects que présente certains mots clé peuvent éventuellement être combinés pour l'élaboration de glossaires que je voudrais d'ailleurs appeler "glossaires combinés".

Je suis parfaitement conscient du caractère pragmatique, <u>non</u> scientifique de cette approche. Mais si un certain nombre de principes de classement sont convenus avec précision, si l'on établit des règles pour empêcher la croissance sauvage des stocks, si l'on réalise sur cette base une division nette des activités, et si les unités terminologiques ne sont classées définitivement que par celui désigné pour le faire, la classification des notions, avec de bonnes possibilités de recherche, paraît certainement possible, y compris au sein des stocks d'autres organisations.

avril 1979

A.A.P. Bothe
Chef du Bureau de Terminologie
et de Documentation
Division des Services linguistiques
Ministère néerlandais des Affaires étrangères

DISCUSSION

In the discussion, chaired by Mr. Felber, the question of experience with data exchange was raised. Mr. Goetschalckx stated that he had been able to incorporate data into his bank with minimal adjustment of the subject classification used. Convertibility of classification schemes was essential. Classification could not be based on single use. Combination of subject codes in search allows much greater precision of retrieval.

- The co-operative work done by the group of which Mr. Bothe spoke is not to be generalized as classifications reflect different retrieval requirements.

- It was generally felt that harmonization of classification schemes was desirable and that classifications used in other areas might provide useful new ideas.

6.5.3 INVITED REPORT 3

M. Krommer-Benz
Infoterm
Vienna, Austria

STATE-OF-THE-ART REPORT
PROJECT "VOCABULARY OF TERMINOLOGICAL DATA BANKS"

1 Purpose and scope of project

The ever increasing interest in and importance of data banks focusing specifically on the storage of terminological rather than linguistic and/or translation data in general is being witnessed by the number of such data banks either in operation already or in the planning stage.

Anybody trying to cope with the literature available on the subject right now will be perplexed at the multiplicity of terms used both in the description of such systems and theoretical treatises. Considering the development of "word bank" to "terminology bank" and, finally (as exemplified by the title of this conference) "terminological data bank" it becomes apparent that something has to be done to ensure unhampered communication at least between those busily working on the problem - namely, between terminologists.

The vocabulary project we have in mind is to focus on concepts and terms relating to recording, processing and dissemination of information of terminological data; thus, this project will not cover concepts and terms on hardware.

Since ISO/TC 37 "Terminology (principles and coordination)" is revising document ISO/R 1087-1969 "Vocabulary of terminology", data already collected and work done so far by this committee are being taken into consideration to avoid duplication of effort. In addition, work on the clarification of concepts and terms in this field is also being undertaken by INTERCONCEPT, the information system for the social sciences at large, and collaboration has been assured with members of INTERCONCEPT as well.

2 Collection of data

2.1 So far Infoterm collected data available on this subject from existing terminological data banks. (A circular to this effect was disseminated last summer.) Thus, glossaries, lists of terms, etc. pertaining to terminology banks have been received amounting to some 300 terms (basically in English and French, some in Swedish as well) to be stored in a card file right now.

2.2 In addition, some literature on the subject has been scanned and terms falling within the scope of the project have been stored in a card file as well. Number of terms collected: approximately 250 (in English, French; some in German, Italian, and Danish).

2.3 Additional data to be collected:

In 1978 Infoterm prepared a bibliography of articles, books, etc. on terminological data banks. It would, therefore, be advisable to scan the entire spectrum of these documents as has been done for a few already (see above, item 2.2).

Cooperation will, of course, be necessary from all to enable us to collect all data in existence right now. Some participants have already expressed their willingness to process the data available so far and/or to supplement these.

3 Proposal for further work

Infoterm will not be able to complete the project on its own; we, therefore, suggest to provide a broad classification scheme for data available in our card file right now. It would, of course, be highly desirable if the data could be stored in one of the terminological data banks so that checking and matching procedures (difficult at the present "manual" stage) could be facilitated. Infoterm is also willing to organize the collection of additional data provided collaborators within TermNet are willing to contribute their share in a format (adopted already by Infoterm for the data collected so far) to be processed easily.

After the data have been collected and classified on the basis of the broad classification system mentioned above proper systematic work will have to be carried out by a working group. Hopefully, results of revised ISO/TC 37 document ISO/R 1087-1969 "Vocabulary of terminology" will be available by then so that some concepts and terms already agreed upon internationally can be utilized for this TermNet project.

DISCUSSION

The chairman, Mr. Felber, asked Ms. Krommer-Benz to report on two new projects, INTERCONCEPT and the Vocabulary of Terminological Data Banks. According to Ms. Krommer-Benz, INTERCONCEPT, a project sponsored by Unesco aims at establishing a terminological data bank to aid the analysis of key concepts and terms in the social sciences. As terminological data banks have to cater to the specific needs of the users, some additional data elements will have to be considered to provide (in this case) social scientists with specific information requested. (Mr. Rigg's paper "A note on citations" gives some preliminary suggestions with reference to INTERCONCEPT's purpose.) A bibliography project, namely "International bibliography of vocabularies and thesauri in the social sciences" undertaken by Infoterm for the specific purpose of the INTERCONCEPT project is to be completed in collaboration with other social science institutions.

It was felt that clarification of concepts and terms used in reference to the operation of terminological data banks was necessary. In co-operation with INTERCONCEPT preliminary work on a "Vocabulary of terminological data banks" had been started last year and was to be continued in collaboration with other TermNet members.

Mr. Goffin gave additional explanations on the project "Vocabulary of terminological data banks". He offered to collaborate in the project.

An offer by GOSSTANDART to supply the Russian dimension to the vocabulary was gratefully received.

6.6 OTHER PAPERS DISTRIBUTED ON THE CONFERENCE

A. Al-Kasimi
Bureau of Coordination in the Arab World
Rabat, Morocco

6.6.1 TOWARDS A CENTRAL TERMINOLOGICAL DATA BANK IN THE ARAB WORLD

It is my pleasure to extend to this distinguished gathering the kind greetings and best wishes of the Director General of the Arab League Educational, Cultural and Scientific Organization (ALECSO), and those of the Director of the Bureau of Coordination of Arabization in the Arab World (BCAAW).

100. Technical Terminology in the Arab World: Coinage and Difficulties:

The creation of scientific and technical terminology in Arabic is undertaken by various institutions, organizations and individuals all over the Arab World such as the Ministries of Education, Arab Academies in Cairo, Baghdad, Damascus and Amman, Universities, lexicographers, translators and authors of scientific books.

At present, the following linguistic means are used in the creation of terminologies in Arabic: 1) borrowing, 2) translation, 3) giving new meanings or extending the meanings of existing words, 4) compounding new words from Arabic and/or other language elements, and 5) coinage.

Unfortunately, there is a great deal of duplication and inconsistency in modern Arabic terminologies. This is due to several factors, the most important of which are the following:

1) Diversity of source languages: English is the source language of some Arabic terminology in the Arab countries of the Middle East (e.g. Egypt, Iraq, Jordan, and Saudi Arabia) whereas French plays that role in the Arab countries of North Africa (e.g. Morocco, Algeria, and Tunis).

2) Linguistic factors: diglossia in Arabic, its numerous dialects, and the richness of its vocabulary, synonymous and polysemous words contribute to the difficulties of the creation of a standardized terminology.

3) Lack of coordination: for a long time, there was no co-operation or coordination among the various organizations in charge of the creation of technical terminology in the Arab world.

200. Coordination and Standardization of Arabic Terminology

Ten years ago, the Arab League charged our Bureau with the task of coordination and standardization of terminologies in the Arab world. The Bureau collects, coordinates, and up-dates the terms coined by various linguistic and educational institutions and publishes them in trilingual specialized glossaries (Arabic-English-French). These glossaries are submitted for approval to the Pan-Arab Conferences on Arabization, which are held periodically by ALECSO to which our Bureau has been attached since 1970. Once they are approved by the Pan-Arab Conferences, the standardized terms become legal and should be used all over the Arab world. The Bureau has already published about 80 trilingual specialized glossaries or dictionaries in various fields of knowledge.

300. Objectives of Arabization

Our objectives are based on the needs of the users of terminologies in the Arab world. Our immediate objective is to provide the higher educational institutions with standardized Arabic equivalents of the English and French terminologies used by them. The majority of Arab universities use English and French as the medium of instruction in their scientific, medical, and technical courses. But they are planning to arabize their medium of instruction soon.

Another major objective is to produce a polytechnic dictionary that facilitates the absorption of Western technology and speeds up scientific and economic cooperation between the Arab Nation and the Western Nations.

400. A Central Terminological Data Bank in the Arab World

It goes without saying that a central terminological data bank in the Arab World will be most useful in the process of storing, coordinating and standardizing terminologies created by various linguistic and scientific institutions in our twenty-one Arab countries. However, you might be surprised to learn that although computers have been utilized fairly widely in the Arab world, these have not yet been utilized in lexicography. One reason for this delay has been that until recently computers were not hospitable enough to accommodate our Arabic script: and transliteration is a very tedious and time-wasting process. So far, our Bureau has been playing the role of a manual central bank. In my opinion, this delay has not been without benefit to us in that it has forced us to acquaint ourselves with the experimentations and experiences of other nations in this field. In the light of previous experiences, we will choose, adapt, or develop the software that corresponds exactly to our objectives.

Arab experts are studying the possibility of creating a central terminological bank which all users - such as Arab Academies, Universities, Ministries of Education, etc. - can contribute to and make use of, through appropriate terminals.

At present some experiments on a limited scale are being undertaken in the Arab World. The University of Mousil in Iraq is storing in its computer scientific and technical terms relevant to a number of courses. The national Moroccan Institute of Research and Studies for Arabization in Rabat has started using a bilingual CRT terminal in storing its terminologies in the data bank of the European Space Agency in Frascati. The Institute, which is aided by UNESCO, ALECSO, and the Moroccan Government hopes that by 1983 it will have stored about a million Arabic terms and their French equivalents. Our Bureau is involved in supporting and supervising this project.

Our Bureau has also reached an agreement with SIEMENS in Munich: we provide the Language Services Department of the Company with our trilingual specialized glossaries and dictionaries to be stored in their data bank. It is hoped that SIEMENS Linguistic program TEAM will be capable shortly of giving the output over a CRT typesetting machine. For the time being and until the central terminological bank is created, the Bureau is considering the installation of a terminal that could facilitate our cooperation with other data banks.

500. Exchange of Arabic Terminology with Data Banks

It is the policy of ALECSO to provide national and international bodies with its educational, cultural, and scientific information freely and without any financial obligations. This is done on the basis of mutual cooperation and cultural exchange. Scientific and technical terminologies fall into this category. Our Bureau is prepared to supply interested data banks with technical glossaries and general dictionaries produced by us or compiled elsewhere in the Arab World. The Bureau is also ready to extend its advice on Arabic linguistic and lexicographical matters when sought. On the other hand, the Bureau hopes that it will benefit by the experience of the international terminological data banks and the recommendations of their first international conference in Vienna concerning quality criteria for the exchange of terminological data; and the principles of creating a terminological bank; so that the process of cooperation with other data banks will be facilitated.

M.A. Dovbenko
VNIIKI
Moscow, USSR

6.6.2 CONTENT AND PRESENTATION OF DATA ELEMENTS FOR TERMINOLOGICAL DATA INTERCHANGE

1. The First International Conference on Terminological Data Banks shows much interest in the problems of content and presentation of data elements for terminological data interchange. In particular, the report by Mr.K.A.Brinkman and Mr.W.H.U.Schewe and TERMNET 5 (5)-79 document "Guidelines for Recording Bibliographical Data" are partially or fully devoted to this problem.

Our point of view on this problem is given below. First of all it is necessary to distinguish presentation of bibliographical description of terminological documents and terminological data preparation for input into the bank of terms. Mr. W.H.U.Schewe's report also contains this distinction. As we represent a standardizing organization we are mainly interested in data processing of terminological standards and recommendations.

We want to note with satisfaction that our approach is shared by AFNOR and DIN.

2. As this conference deals primarily with the problems of establishing network for bibliographical data interchange on terminology, first we shall touch the state

of this problem in the USSR and our opinion on the content of data elements for bibliographical description and TERMNET interchange data format.

Last year State Standard GOST 7.14-78 "Communication Format. Structure of Bibliographic Record and Data Elements on Magnetic Tape" was adopted; this standard is in complete conformity with International Standard ISO 2709. March this year State Standard on content of communication format regulating standardized content and presentation of data elements on magnetic tape was also adopted. GOST 7.14-78 will come into action beginning with January 1, 1980. These standards regulate data recording of books, serial publications, standards, patents and other kinds of documents.

These standards secure information compatibility with ISONET.

While preparing for the conference we carried out a comparative analysis of TERMNET and ISONET formats based on "TERMNET 5 (5)-79" and "ISONET manual - 78"; it is neccessary to point out that much work has been done on selecting data elements for bibliographical data interchange in terminology and coordinating TERMNET worksheet with that of ISONET. But in our opinion the coordination has not been carried out to the end:

1) contents of mandatory data elements of ISONET and TERMNET (DT 2) formats do not coinside, in particular in TERMNET format there is no such an important data element as "record status" (new, replacement, removal);

2) sometimes presentation of data elements differs in these formats (as in ISONET format dates are presented in a unified form according to ISO 2014, and in TERMNET format data elements with labels 003, 101, 205, 206 and 209 are presented in three different forms which we think is hardly reasonable; ISONET format recommends to use two-letter codes for languages (label 300) while TERMNET format uses different ones; data elements with labels 304, 500 and 501 are treated differently in these formats).

We think it advisable to complete document TERMNET 5 (5)-79 in order to provide full compatibility of the mandatory data elements in appendix 2 to this document with ISONET format. For this purpose it is useful to establish working contacts with INFCO/WG 1.

After eliminating the above-said differences in mandatory data elements our institute is ready to cooperate with TERMNET by giving bibliographical information on state and branch terminological standards on magnetic tape or microfiche and receiving information on standardized vocabularies of other countries from TERMNET in exchange.

3. As far as automated processing of standardized (recommended) terms is concerned the input into our data bank is restricted to the following data elements for each term:

1) record status;
2) designation of a document specifying a term;
3) date of approval (publication);
4) term number in a document;
5) main entry;

6) admissable synonym;

7) abridged form;

8) depricated synonym;

9) term equivalents in foreign languages;

10) term definition in Russian.

The given set of data elements differs from that used in other systems, for example, in "TERM" (FRG) or "TERMDOK" (Sweden) as it does not contain any grammatical information on a term. This is the result of the fact that State Standards, ISO, CMEA standards do not contain this kind of information.

Information on machine-readable carriers does not differentiate capital and small letters, diacritics are eliminated but for typical replacements: ä - ae, ü - ue, õ - oe. Indiscrimination of capital and small letters is due to the characteristics of standardized peripherals of EC computer in which the set of characters contains only capital letters of Latin and Cyrillic alphabets besides special marks and figures.

After amending DP 6156 format in acordance with our remarks we are ready to cooperate in terminological data interchange with other banks of terms within the limits of the above-mentioned restrictions.

А С И Т О

Форма Т1

Элемент данных	Метка	Содержание
Код формы и статус записи	00=	Т1 (N-новая, С - измен., D - иск
Обозначение документа	01=	
Дата утверждения (издания)	02=	(формат ГГММДД)
Код по КГС	03=	
Номер термина в документе	09=	
Заглавный термин	10=	
Допустимый синоним	11=	
Краткая форма	12=	
Недопустимый синоним	13=	
Эквивалент на немецком языке	20=	
Эквивалент на английском языке	21=	
Эквивалент на французском языке	22=	
Определение на русском языке	50=	

1. Заполнил________________________ Дата________________________

2. Проверил________________________ Дата________________________

3. Подготовка на МЛ____________ Дата________________________

M.A. Dovbenko
VNIIKI
Moscow, USSR

6.6.3 GOSSTANDART ACTIVITIES IN TERMINOLOGY AND TERMINOLOGICAL DATA BANKS INTERACTION

The view expressed in this report reflects the terminological problems faced by USSR State Committee for standards (Gosstandart).

On the national level the tasks of Gosstandart are as follows:

- standardization of scientific and technical terminologies for the needs of national economy of the USSR;
- control over the use of standardized terminology in all kinds of standards and standardized documentation;
- providing specialists with information on standardized terminologies.

VNIIKI is the scientific and methodological centre on standardization of terminologies within the organizational system of Gosstandart.

On the regional level (within the framework of the Council for Mutual Economic Assistance) Gosstandart participates in the development of CMEA terminological standards and CMEA Guides for standardization.

On the international level Gosstandart, being the Soviet Member of ISO, International Electrotechnical Commission and European Organization for Quality Control responsible for standardization and unification of terminologies, participates in the preparation of ISO terminological standards, the International Electrotechnical Vocabulary as well as EOQC Vocabulary.

In its international efforts Gosstandart also carries out the work of methodological character holding the Secretariat of ISO/TC 37/WG 1 "Principles of Terminology".

The above-mentioned tasks determine the approach which may differ if we take into account some other Soviet organizations working on terminologies, for example, the USSR Academy of Sciences, universities.

1. At present, computerized terminological data banks are being developed on a large scale throughout the world. The main tasks of such banks are collecting and up-dating terminological data files used for the preparation of various terminological vocabularies, indexes and lists of terms and documents, and providing assistance to translators of scientific and technical literature.

When analysing terminological data banks it is useful to emphasize the character of the information used and their scope of application. Taking into account the tasks faced, Gosstandart is mostly interested in automated banks of terms. One should not think that we are not ready to exchange terminological bibliographies. We are just stressing the priorities As far as terminological bibliographies are concerned bibliographies of standardized vocabularies are of primary importance to our organization. In respect to the character of information stored the terminological data banks are divided into those that contain only standardized terminologies and those that include non-standardized terminologies taken from different guides and vocabularies as well.

The goals of Gosstandart and our institute in the nearest future are mainly connected with the bank of standardized terminology.

The main fields of application of terminological data banks include preparation of terminological indexes and vocabularies and translation of scientific and technical literature.

We think that the most effective application of terminological data banks is computer-assisted photo-setting of vocabularies and indexes of terms.

2. It has become obvious that the current progress in computerization of terminological information brings to life the necessity of uniting terminological data banks in an information network analogous to INIS, MEDLARS and other information networks.

It is evident that effective interaction of the network banks necessitates preliminary work on compatibility of information and terminological systems and solving a number of organizational and legal problems.

The notion of information systems compatibility is rather wide, and includes technical and information compatibility. Technical compatibility is achieved by using unified (compatible) technical means, data carriers, and machine-readable record characteristics. At the modern state of computing machinery these requirements can be easily met in the existing computerized data networks by using the standard 1/2 inch magnetic tape for data exchange.

Information compatibility of terminological systems presents greater problems due to the following:

1) absence of standards on the structure and content of communication format for terminological information;

2) absence of a unified information retrieval language of a classification or thesaurus type;

3) multilingual character of information.

We think it abvisable to develop a communication format for terminological data exchange on the basis of ISO 2709 on the structure of a communication format.

The set of data elements given in ISO/DP 6156 can be taken as the basis for the development of the content of a communication format. Among the most important items of this work there is the choice of mandatory data elements and securing language equality of the format content for usually terminological information is of multilingual character.

The incompatibility of national classification systems can be overcome only with the help of a unified classification system for scientific and technical terminology using not more than 3 or 4 levels.

The ISO Thesaurus can be used for indexing of bibliographic data on terminology.

Efficient international exchange of terminological information and standardized terminology in particular between countries, speaking various languages, can be reached only by joint efforts in its development.

3. The Gosstandart Automated Terminological Information System is a bank of standardized (recommended) terminology from USSR state and branch standards, ISO and CMEA standards, recommendations of the Committee of Scientific and Technical Terminology of the USSR Academy of Sciencies and IEC publications. The bank is operated on EC-1022 computer with immediate-access memory of 512 Kbt and removable disk packs of 29 Mbt; the bank contains about 60 thousand records (vocabulary entries).

Each record includes a term, its abbreviated form, synonims, equivalents in foreign languages (English, German and French, and for CMEA standards Czech, Polish, Rumanian, Bulgarian and Hungarian), a definition and number of auxiliary data elements (designation of the standard-type document specifying the term, the date of the document approval and publication; document code according to the USSR State standards classifier, etc.).

The system enables to prepare terminological collections and term indexes (subject and alphabetic, permutational, numerical). Besides, information retrieval and up-dating of data base in an interactive mode with the use of alphanumerical local videoterminals are available.

The structure of the communication format used is in full agreement with ISO 2709.

Now we can make the following conclusions:

1. Gosstandart supports the idea of a network of terminological centres with automated banks of standardized terms.

2. For the implementation of the network and coordination of its functioning the following is of primary importance:

a) development and approval of ISO standard on the structure and content of a communication format for terminological information;

b) creation of a unified (or a series of compatible) classification for terminological data exchange.

In conclusion we would like to express the hope that the work begun will widen scientific and business contacts between the countries participating in TERMNET project.

Food and Agricultural Organization
of the United Nations, FAO
Rome, Italy

6.6.4 INTERNATIONAL COOPERATION BETWEEN TERMINOLOGICAL DATA BANKS. VIEWS AND EXPECTATIONS OF THE ORGANIZATIONS OF THE UNITED NATIONS SYSTEM.

1 Exchange of Terminological Data

At the last Interagency Meeting on Language Arrangements in New York, September 1978, it was confirmed that the exchange of terminological data between organizations should be initiated by the gradual introduction of EDP techniques, with the possible establishment of a terminology data bank, supporting United Nations needs, as its ultimate goal. As was stated, the exchange of terminological data within the international community not only increases the availability of terminological information, but also avoids duplication in the collection and dissemination of terminological data.

The general political trend in the world today is toward giving ever more importance to languages, stimulated by the emerging nations' search for identity with the passing of the colonial era. This development makes international collaboration in the field of terminology even more imperative. Therefore, the standardization of terminology and the observance of already existing international standards presented by ISO/TC37 - "Terminology (principles and co-ordination)"is something that must be given high priority.

One of the fundamental problems in the exchange of terminologica data through EDP techniques was to decide on a format which would facilitate data exchange. Such a format has been elaborated in ISO/DP 6156 MATER (Magnetic Tape Exchange for Terminological/Lexicographical Records). It is now necessary to find and agree on a harmonized or uniform classification system to be used by all terminological data banks in the exchange of terminological data. Why the UDC (Universal Decimal Classification) is excluded by many terminology experts, is not clear. I should like to express the opinion that it would be better to work with such a world-wide and well-known classification as the UDC, than under the present conditions, e.g., each terminology data bank having a classification system of its own, thus obliging the users (translators, revisers, interpreters, editors, etc.) to study first the system of the data supplier.

It is hoped that this problem may be solved during the Conference by either finding and agreeing on a new uniform classification system, or by using the UDC.

Some other basic problems which may be solved during the Conference regard the

2 <u>Guidelines for the Establishment of New Terminological Data Banks</u>.

These problems concern:

a) the compatibility of input data based on an accorded input sheet;

b) compatible transliteration systems for non-Latin languages, a most important item, especially for the organizations of the United Nations system with multilingual terminology including Arabic, Chinese, and Russian; the transliteration systems must be compatible, in order to make the exchange of multilingual terminology between data banks possible;

c) quality of terminology, which is essential, since experience in EDP storage has shown that rectification of carelessly selected terminological data is time-consuming as well as wasteful of financial resources; therefore levels of reliability should be established and coded on the input sheet.

It may, furthermore, be stated in the Guidelines that all specialized institutions and organizations with established terminological data banks /here understood as: 'collections of terminology, without regard to their technical realization'/ retain authority for the terminology in their special subject fields; so FAO's terminology concerning agriculture and related subject matters should be authoritative in that subject field, as should WMO for meteorology, WHO for health, Unesco for education, science, etc.

R. Goffin
Commission of the European Communities (CEC)
Brussels, Belgium

6.6.5 THE VOCABULARY OF TERMINOLOGICAL DATA BANKS

Working Document

It is not easy to delimit the special vocabulary of terminological data banks, because it comes from a field of study which is, by its very nature, interdisciplinary : applied computerized linguistics. If a little simplification be permitted, we might say, in this respect, that computer techniques are used to assist linguistic research and more especially terminological research.

This vocabulary is drawn from the following specialized languages : lexicology and lexicography, terminology, documentology, automatic non-digital data processing (software and hardware).

The vocabulary of the automatic processing of terminology - computerized terminology - should therefore include only terms specific to that part of terminological work which is computerized. This field is a very narrow one. Though computer procedures are not, in general, used during the collection and collation phases of terminological work, they are used for "Question and answer" interrogation, batch interrogation, the automatic printing of glossaries and of computer listings with alphabetical indexes, and the collection of statistics by subject field, by language and by terminology user. Moreover, as the terminological cards are prepared with a view to processing them by computer and according to computerized specifications or constraints, the vocabulary linked to terminological work prior to the computer phase can reasonably be held to be a part of this vocabulary.

There are four headings to be included in the list :

1. Basic terms
2. Terminological research prior to the computer phase
3. Description of the terminological card
4. Interrogation and consultation at the terminal and at the printer.

It is to be noted that each terminology bank has evolved a jargon of its own for the description of its particular activities: the esoteric nature of the language is in conflict with the need of each bank to reach beyond its own resources and with the need to facilitate exchanges. The cryptic nature of this jargon is the result of the diversity of the hardware, the degree of specificity of the terminological needs of each unit and the near impossibility of standardizing the systems. Moreover, this vocabulary is characterized by an unparalleled explosion of neologisms and the often chaotic profusion of synonyms. That is why this working paper is only a first attempt to collect this fledgling terminology.

M. T. KHAFAGY
The Arab Educational, Cultural
and Scientific Organization (ALECSO)
Cairo, Egypt

6.6.6 ALECSO AND ITS EFFORTS AT UNIFYING ARABIC TERMINOLOGY AND THE STANDARDIZATION OF ITS APPLICATION

Introduction: The Problem of Arabic Terminology

The problems of Arabic terminology differ from those of other languages. This is due to the evolution of Arabic throughout its long history as well as the social influences affecting those speaking the language. Arabic is a language that enjoys a wide range of synonymous words though many linguists tend to challenge this fact. This synonymy could be due to the diversity of Arabic dialects, its widespread use and contact with many other languages using many of the latter's synonyms.

It goes without saying that the development curve and flourishing of a scientific term is closely linked to the scientific activity of the linguistic community. Arabic played the role of the universal language of human knowledge over a long period of time, through it Moslems contributed to the enrichment of that knowledge and it was the language that conveyed to the West the Greek and Syriac tomes in medicine and mathematics and other sciences that fall today under the name of pure and applied sciences. But the link binding those speaking Arabic with scientific development and its subsistence was severed for over four centuries leading to the widening of the lexical gap between it and other languages that kept pace with scientific development such as German, English, French, and others.

With the emergence of the modern Arab renaissance at the turn of the 19th century and the revival of Arabic as a language capable of absorbing the progress of the era and fulfilling the needs of those speaking it in expressing their thoughts in all fields, translation and Arabization were propelled forward. Nevertheless, the vocabulary obtained from those efforts were mere derivatives until the establishment of Arabic language scientific academies at the start of this century. This dispersal was a load on the language resulting in contradiction and confusion in the selection of Arabic synonyms of foreign terms.

At present, it is possible to divide Arabic terminologies into two categories:

a. Terminology or terms related to "concepts" which have their roots in Arab heritage.
b. Terminology related to new ideas or borrowed from other cultures.

The terms falling under the first category are, in general, of Arabic origin and the margin of contradiction in content and form are very rare. Even the terms that have been Arabized have settled, as they have acquired phonetic and grammatical characteristics very near to Arab understanding.

As for the terms of the second category, they suffer from two sets of factors: one related to the Arabic language itself; and the second to those speaking Arabic.

As to the factors related to the Arabic language, they are concentrated in the language's flexibility and the various alternatives available for the expression of one idea. This is obvious in the abundance of synonyms and the flexibility of morphology laws especially those related to verbal nouns.

The factors related to those speaking Arabic are numerous for there are those connected with social upbringing conditions, those related to the linguistic and subjective capabilities and the extent of understanding the idea correctly.

The Arab World does not enjoy total linguistic harmony as one may think, although Arabic is the official language. For in addition to the multiple languages that are used, we find linguistic duplicity in Arabic itself, e.g. classical Arabic and dialects. This resulted in the weakening of the linguistic intuition of those speaking the language. Add to this the lack of strong links with Arab heritage, for a large portion of it is yet to be published while the part that has been published lacks the minimum organizational services such as indices and word lists.

A scientific term cannot develop from nothing only its application can guarantee its spread and flourishment. A number of Arab specialists are using other languages such as English and French with the aim of enlarging their contacts. Arab linguistic academies have not, unfortunately, realized this fact as they still treat the term as an aim in itself giving attention to dictionaries and not other means of propagation such as the actual application of the term whether on the official level (in the names of organizations and institutions) or in specialized thought production. In addition, linguistic academies and societies have not yet found a systematic method to collect specialized terminologies and prepare them for those in charge of selecting their Arabic synonyms. They have also brought up linguistic problems that have led many specialists to refrain from participating in the Arabization process and in enriching Arabic with terms and words.

All these factors combined make the problems of unification and standardization of terminologies on the pan-Arab level a national objective so as to encourage the application of Arabic in the different scientific fields.

Unification ot Arabic Terminology

The Arab States suffered, and are still suffering, from the problem of the application of scientific terminology and its transfer into Arabic in a unified manner capable of encompassing the different sciences. One of the main reasons for this problem was the numerous cultures that dominated the Arab region during the past two centuries. Efforts were exerted in translation and scientific writing but the persons undertaking this operation either transliterated or selected the nearest Arabic equivalent and applied it. This selection, of course, was governed by the extent of the selector's education, culture, and specialization and was affected by the efforts of scientific and linguistic academies in the area. If we add to this the multitude of languages which are translated from and which are used as references in the preparation of researches (at least seven languages) we can visualize, in general, the problem of using terminology in the Arab region. It is noteworthy that the question of applying standards occupied a major part of this problem, as Arabic standards and rules were introduced to the region under the factors mentioned previously. This led to two main results: one was that a large portion of these standards were translated into Arabic several times over by Arab specialists leading to confusion, duplicity and diversity; the second result was that this situation slowed down the process of the application of standards. As a result of this, the Arab scientific institutions found themselves in a difficult situation faced with a massive influx of new standards. This situation was further aggravated with the passing of time which in turn deprived the Arab States of the continuous exchange of information with international specialized organizations.

ALECSO's Efforts in the Field of Terminology

There are two main organs affiliated to ALECSO concerned with terminology in the region: the Arabization Coordination Bureau (Rabat, Morocco) and the Information Department.

The Arabization Coordination Bureau, established in 1961, was affiliated to the Secretariat General of the League of

Arab States until ALECSO came into being in 1970. At the start, the Bureau's aim was to Arabize foreign terminologies in the different sciences and propagate them among the Arab States. To achieve this, it published several lexicons covering various sciences. These lexicons included scientific and technical terminology translated into Arabic by Arab linguistic academies, universities and scientific councils. It also activated translation in the Arab World in general and the Arab Maghreb in particular.

Lately (after being affiliated to ALECSO), the Bureau's work has spread to include the application of Arabic in teaching at all educational levels; convening Pan-Arab Arabization conferences; publishing a greater number of lexicons; coordinating Arab efforts in enriching the Arabic language with terminologies; and unifying scientific terminology in the Arab States. The Bureau has published a large number of lexicons dealing with chemistry, jurisdiction, information sciences, mathematics, oil, geology, plants, insects, statistics, etc.. We cannot, in all honesty, say that these lexicons are comprehensive, but they have provided the Arab States with the tools; tools that can be modified continuously, and have laid down the rules and bases to be used in achieving this end.

The Information Department (ALECSO) since its establishment emphasized in its projects and programmes the problem of terminology. Through meetings and conferences (which started in Damascus in 1971), the Arab World's need for unified efforts in compiling terminology in general, and that of information sciences in particular, was brought into being. This was in compliance to the dire and true need of scientific research organizations and research workers in the Arab States, this need that was portrayed in the meetings and conferences and the questionnaires prepared for that purpose.

The past, present and future efforts of the Information Department can be summarized as follows:

a. <u>In the field of statistics:</u> This is a vital task undertaken by the Information Department. It has put down Arab statistical terminology that is in line with international standards and that faces the needs of the Arab region. This terminology was discussed at a meeting convened for this specific purpose in Baghdad in the Spring of 1977. The terminology was used in the statistical tables sent out annually by ALECSO to the Arab States and which form the bases of the Arab statistical bulletin published every year. It must be stated here that new problems have arisen in this field especially after the wide-spread use of the statistical sheets prepared by ALECSO and which include collecting statistics related to the different stages of education or related to a certain field selected by ALECSO for statistical analysis every two years. These statistical sheets emphasized the question of terminology, as each included a list of the terms used and their explanations. An Arab meeting is foreseen in ALECSO's 1980/1981 programmes to further discuss statistical terminology. It may be added that since 1978 ALECSO has been preparing for the establishment of a statistical data base. This led the Information Department to penetrate deeply into statistical terminology processes whether they be concerned with statistical sheets or applied in the building up of the system. After the completion (at the end of 1981) of the thesauri related to this topic a valuable tool for work will be made available to the Arab Region.

According to the set plan, it is expected over the next six years to augment the acquisition of statistical terminologies available to ALECSO due to the comprehensiveness of these statistics. At present, the statistical terminology covers adequately the field of education but lags in the fields of culture and science as their related statistical sheets are limited in quantity and quality. This is due to reasons related to the manpower and material capabilities of the Region and the competence of statistical systems in the Arab states and their response to the requirements of the region.

The problems facing ALECSO in the field of statistics are:

1. The statistical agencies in the Arab States have not yet complied sufficiently with the terminology included in the sheets provided by ALECSO.
2. There is no set system in the Arab States in choosing the Arabic equivalent of the foreign term.
3. The difference in the structure and systems covered by the terms makes it difficult to generalize them in all the Arab Region.

b. In the field of bibliography: It must be stated that ALECSO's Information Department is involved at present, and in future plans, in compiling specific and unified terminology in the Arab Region. Due to the difficulty of this task and to the lack of awareness, the efforts exerted in this respect at the beginning were limited and time-consuming. It is still in the early stages although the rate of development is speeding up. The Information Department is working within set principles reached as a result of analysing the actual Arab status vis-a-vi the international situation. These principles are:

1. The continuous international efforts at standardization, formulation, and application of terminologies.

2. Recent attempts aimed at developing traditional information operations have resulted in a somewhat confused application method especially after the use of the computer in information and bibliographic operations.

3. This new situation creates a difficult task apropos the Arabic language which is trying to assimilate new terminologies in such a manner as to be capable of transferring its correct content to the Arab user.

4. The application or use of terminologies within the framework of the Information Department's interests makes the latter's activities closely connected with the terminologies of the other branches of science and culture.

If we review briefly what has been accomplished in this field, we can say that the Information Department discussed at Arab conferences and seminars the problems of terminology and unification and brought to the attention of information personnel the need for Arabic: classification schemes, subject heading lists, standards for bibliographic description, unified entries of Arab authors, and an Arabic thesaurus in the fields of education, culture and science. These stages took several years and ALECSO managed to bring into being the following:

A. The Arabized Standard for Bibliographic Description in which the rules were translated and foreign examples were substituted by Arabic ones. Three issues of it were published in compliance with the modifications that these standards were subjected to and the second issue included a complete list of the Arabized terminologies. This document evoked widespread discussion in the Arab Region as a result of which ALECSO published number four in its series on information studies that dealt with the different view points. What is of importance to us is that the second issue contained a large number of terms amounting to around 300 along with their definitions.

The Information Department devoted all its efforts and capabilities (newsletters, periodicals, books, conferences, seminars, etc.) to promote awareness and importance of the terminologies included in ALECSO's works and that they be depended upon and used, after discussion and modification, so as to be in compliance with the present needs.

B. The Arabic Subject Heading List

The lists prepared by ALECSO included terminology used in the Arab States and compiled by experts from countries with outstanding activities in the different fields of knowledge. It was taken into consideration that the Arabic term should and must serve Arabic indexing operations. These lists were provided with definitions explaining the meanings of the terms and how to apply them as accurately as possible.

C. The Arabic Classification Scheme and Subject Bibliography

This continuous programme is in accordance with the recommendations passed by several Arab conferences. Several parts of the scheme related to Islamic disciplines and education have been prepared while the section related to social sciences is at present being compiled. The Arabic classification scheme is going hand in hand with the subject heading lists and they are rich with Arabic and Arabized terminologies that are derived from Arab linguistic and scientific academies.

D. Arab Studies in the Field of Information

The main aim of these studies was to provide scientific material in Arabic in the field of information. The seven published studies included new terminologies introduced to the Arab reader for the first time. However, it should be stated that due to the wide range of topics treated by these studies and the dive writers who compiled them, they have not yet made the required impact vis-a-vis terminology and although they are used as text-books by some Arab universities, this is but the first step.

E. Arabic Thesauri

To provide the necessary tools for non-traditional retrieval systems, the Information Department is compiling a number of Arabic specialized thesauri to be used in the subjective analysis of Arabic and non-Arabic documents. These thesauri are considered, first and foremost as standardization and unification of terminology.

F. The Use of Computer in Bibliographic Activities

The use of computer in bibliographic activities has become a must as it achieves not only flexibility, precision and speed but also avails statistical data that serves in the study of the structural characteristics of literature. Taking this vanguard role into consideration, the Information Department of ALECSO began to lay down a system for the establishment of an Arab national bibliographic control centre that covers

Arab national bibliographic records, the latter hoped to be one of the components of the Universal Bibliographic Control (UBC) system. In laying down this system, linkage and compatibility between other national systems was taken into consideration. As mentioned before, the Department is also laying down a retrieval system that includes statistical data related to educational, cultural and scientific activities in the Arab World. Arabic, of course, will be the basic language in these systems.

It is noteworthy, that the use of computer in the analysis of Arabic texts provokes a number of linguistic problems whether in orthography, grammar, semantics or other matters related to the characteristics of the language.

ALECSO has made studies on some of these problems and has reached favourable results therein and encouraged the application of these results in all the Arab World. Foremost among these problems is that dealing with the elements of the Arabic writing system and the full representation of these elements in a computer system. ALECSO is encouraging companies to produce computer hardware capable of assimilating all these elements, especially vocalization which if neglected leads to homographs in Arabic texts.

ALECSO is also encouraging feasibility studies pertaining to the use of computer in sophisticated retrieval systems and capable of handling Arabic texts and identifying the competence of the Arabic language in such systems whether it be related to morphological or syntactical or semantic characteristics of the language.

Furthermore, ALECSO is cooperating with a team of research workers at Cairo University. A member of this team (Dr. Hishmat Kasem) has conducted a comparative study on the competence of Arabic with that of English in some retrieval techniques that rely upon the natural language. This comparison was based upon the two major measures that are used in this field, namely: recall and precision.

M. Krommer-Benz
Infoterm
Vienna, Austria

6.6.7 TERMNET AND INTERCONCEPT

1 INTRODUCTION

In view of the fact that details on Unesco's INTERCONCEPT projects as such are dealt with in a number of other papers presented at the COCTA panel session this paper will focus on INTERCONCEPT in terms of a network for conceptual analysis in the social sciences, and its relation to TermNet, the world-wide terminology network developed by the International Information Centre for Terminology (Infoterm).

2 INFOTERM

Progress in all fields of knowledge is heavily dependent upon the communication of information; unabiguous information, however, is only possible if the concepts have the same meaning for all who participate in the communication process, whether the exchange of information be national or international [1].

In order to meet this "information challenge" it will be necessary, as stated in the "Study Report on the Feasibility of a World Science Information System (UNISIST)", initiated by Unesco, that "the attention of scientists, learned societies and information science associations should be drawn to the need for joint efforts in developing better tools for the control and conversion of natural and indexing languages in science and technology....." [2]

Originally conceived as an information system geared to the natural sciences and technology, the UNISIST Programme, now called the General Information Programme of Unesco, has been extended to include the "soft" sciences and the humanities as well.

It is of particular interest to note that the Advisory Committee of the General Information Programme recently recommended a high priority for terminological work. Thus, both, Infoterm's world-wide network of terminological agencies covering all fields and all languages, and the INTERCONCEPT-Project, an international information system for the analysis and storage of key concepts and terms in the social sciences, represent decisive steps forward towards a goal, still far away, namely international unification of concepts and terms. This, nevertheless, is a linguistic prerequisite, particularly for interdisciplinary collaboration, and also seems singularly appropriate within the larger framework of the world-wide General Information Programme mentioned above. Information networks of a multilingual nature especially, such as those in question, require more sophisticated communication tools based on reliable terminologies which in turn facilitate the preparation of thesauri [3].

2.1 History

The need for an international clearinghouse for terminology has long been felt. However, the establishment of such a centre could not be realized until 1971, when, via contract between Unesco and the Austrian Standards Institute (ON), Infoterm was finally established. (A detailed history is given in Infoterm Series 1. See below).

2.2 Tasks

In compliance with UNISIST Recommendations 4 and 12, Infoterm's functions of coordinating terminological activities carried out throughout the world with a view towards development of a network of terminological agencies (TermNet).

The main tasks are:

- Coordination and advice in the field of terminology particularly with respect to the application of terminological principles and other relevant standards pertaining to terminological documentation
- co-operation with all terminological agencies on the national and international level
- collection and analysis of terminological publications
- preparation of bibliographies of standardized and specialized vocabularies
- preparation of an inventory of all terminology preparing and terminology documentation bodies
- investigation into the possibilities of effective co-operation between terminology banks.

Due to the fact that Infoterm's staff is very small the tasks listed above could not be undertaken simultaneously. Priorities have had to be set and - in agreement with the concept of decentralization in information and documentation - specialized work will have to be delegated.

2.3 Publications

The projects completed so far have been published within the Infoterm Series. These are:

WÜSTER, E. The Road to Infoterm. Two reports prepared on behalf of Unesco. Inventory of Sources of Scientific and Technical Terminology. A Plan for Establishing an International Information Centre (Clearinghouse) for Terminology. München: Verlag Dokumentation, 1974, 141 p., A5. Infoterm Series 1.

FELBER, H.; KROMMER-BENZ M.; MANU, A. International Bibliography of Standardized Vocabularies / Bibliographie internationale de vocabulaires normalisés / Internationale Bibliographie der Normwörterbücher. München: K.G.Saur Verlag, 1979, XXIV + 540 p., A4. Infoterm Series 2.

INFOTERM. First Infoterm Symposium. International Co-operation in Terminology / Premier Symposium d'Infoterm. Coopération internationale en terminologie. München: Verlag Dokumentation, 1976, 332 p., A5. Infoterm Series 3.

KROMMER-BENZ, M. World Guide to Terminological Activities / Guide mondial des activités terminologiques. München: Verlag Dokumentation, 1977, 311 p., A5. Infoterm Series 4.

In addition the Infoterm "Newsletter", published since 1976 as part of "Lebende Sprachen", carries information about ongoing terminological activities and projects.

2.4 Network of terminological agencies

Collaboration of all those involved in terminology work will be essential to avoid duplication of effort, and to ensure closer co-operation between terminologists, linguists and information scientists.

At the First Infoterm Symposium on "International co-operation in terminology", held in Vienna in 1975 it was recommended that Infoterm take immediately appropriate action with a view towards the establishment of a world-wide network for the coordination of terminology work following a detailed study to be carried out in consultation with competent authorities in this field.

This study [4] was then submitted for evaluation to a panel of experts convened by Unesco in Vienna, Oct. 1977 and subsequently redrafted in order to reflect the opinion presented by this group of experts [5].

In compliance with a recommendation passed by this panel, Infoterm has meanwhile proceeded with the implementation of TermNet by sending a communication (inclusive of a questionnaire) and other pertinent documents to a selected number of likely members inviting them to join TermNet. Analysis of the responses has been completed, and a meeting of the proposed advisory panel was held in spring of 1979 to deliberate necessary adjustments, changes and control of such a network. It is expected that with the consolidation of the network a user's committee will eventually take over most of the functions of the advisory panel [6]. A preliminary meeting of prospective TermNet partners will also be convened; whether TermNet will consist of members or partners will depend on the decisions taken by those willing to co-operate within TermNet.

2.4.1 Aims of TermNet

TermNet will function as an individual unit (sub-system) in the General Information Programme of Unesco.
TermNet is intended to encourage the following activities:

- co-operative action in the preparation of terminologies, i.e. division of labour for the preparation of terminologies among expert groups and competent professional organizations, and exchange of terminological results;

- co-operative action in documentation of terminology, i.e. division of labour in the collection and analysis of terminological literature, and recording of bibliographical data, and the exchange of results. The terminology banks are also included here;

- to prepare by a co-operative action of all those joining TermNet in conventional or machine-readable form international bibliographies of vocabularies and the like pertaining to specific fields;

- to collect and/or record in conventional or machine-readable form by a co-operative action of all those joining TermNet all data on terminology work and training;

- to create an interconnexion for the collection and recording of terminological data in conventional (card file) or machine-readable form (automated terminology banks) by a co-operation of all those joining TermNet;

All data thus collected should be exchanged with TermNet members [4].

2.4.2 Members of TermNet

The network should include the following members:

2.4.2.1 Infoterm

Infoterm is intended to function as international focal point for coordination by disseminating information received, by acting as a documentation centre and advising on coordination of efforts; it will also assume responsibility with regard to linkage to other international organizations.

2.4.2.2 International and national specialized terminological centres

These centres function as terminological foci for certain disciplines, such as electrical engineering (International Electrotechnical Commission (IEC)), welding (International Institute for Welding (IIW)), meteorology (World Meteorological Organization (WMO)), law and administration (Legal and Administrative Terminology (LEAT)), etc.

2.4.2.3 Regional sub-networks

A number of regional sub-networks are already in existence, such as NORDTERM (close collaboration between Denmark, Finland, Norway, and Sweden) or the Conseil international de la langue française (CILF) which comprises some 24 French speaking countries. HISPANOTERM, established recently in Madrid, is to coordinate terminological activities within Spanish-speaking countries.

2.4.2.4 National terminology centres

These centres aim at coordination of such activities on a national level. Examples for these are: the Tekniska Nomenklaturcentralen (TNC) in Sweden, the Association française de terminologie (AFTERM) in France, etc.

2.4.2.5 Terminology Bank

Co-operation among existing terminology banks has already been started. Terminology banks either for standardized and/or non-standardized terminological data exist in France (Normaterm), in the USSR (ASITO), within the Commission of the European Communities (EURODICAUTOM), the Banque de terminologie de Québec and the Direction générale de la terminologie et de la documentation du Bureau des traductions,

Secrétariat d'Etat du Canada, etc.

(For a complete survey of existing terminology banks and those in the planning stage, for details on terminological data stored, equipment used, etc. see Infoterm Series 4. World Guide to Terminological Activities).

INTERCONCEPT, together with other sub-networks, such as EURONET, ISONET (see p.8) etc. all of which go beyond the scope of a terminology bank per se would fit into the latter category of prospective TermNet members.

2.4.3 Implementation of TermNet

As mentioned above (item 2.4) the actual implementation of TermNet will commence after detailed analysis of the questionnaires already received; new members are likely to join in the near future. From the answers received so far it can be deduced that a desire and eagerness for closer collaboration and exchange of terminological data indeed exists. As to the structure of TermNet it should be stressed that it would be a flexible and open-ended system so as to cope with and adjust to present and future needs.

2.4.4 TermNet and INTERCONCEPT

Close contacts between Infoterm and INTERCONCEPT were strengthened recently due to the fact that Infoterm has been invited to serve as member of the international INTERCONCEPT Advisory Committee. This is of particular advantage as INTERCONCEPT can thus keep abreast of any and all developments within the newly launched TermNet with particular emphasis on terminology banks.

Chapter 6 of the Infoterm Study on the development of a network for terminology information and documentation [4] also gives an indication of the demands and restraints future TermNet members will have to accept in order to maximize co-operative results. With respect to INTERCONCEPT as a computerized information system, this implies the application of international standards particularly with reference to data input and exchange formats in order to ensure compatibility with other international or national terminology centres or agencies as stressed by Felber [7] and the authors of one of the background papers for the meeting on INTERCONCEPT, in May 1977 [8]. With a great number of these relevant standards being under revision right now, it is essential for future members of TermNet, such as INTERCONCEPT, to be kept informed about these developments as has been done already with reference to record structure, etc. for INTERCONCEPT pilot projects. This is also possible because of Infoterm's close collaboration with ISO/TC 37 "Terminology (principles and co-ordination)" and the latter's close collaboration with ISO/TC 46 "Documentation".

3 SERVICES TERMNET - INTERCONCEPT

The organigrammes drawn up on the basis of discussions and experiences gained so far (such as the exchange of data between the terminology banks of France and Canada respectively) reflects a loose and decentralized structure, both for terminology work in general and for specific fields in particular. It is thus imagined that users should be able to consult whatever terminology service is closer to them. For this reason it is necessary that Infoterm should have at its disposal a properly working information system fed by the members of the system. E.g. a social scientist could thus either address his/her query to Infoterm in case specific information about certain concepts is completely unknown. Infoterm could then refer this prospective user of the network to the appropriate source of information which could be a sub-net of TermNet (such as the INTERCONCEPT data file at Unesco headquarters), an international or national terminology centre or a bank or whatever agency would be most convenient. Of course, problems of payment for services, particularly with respect to data exchange, will have to be dealt with as TermNet evolves.

4 OUTLOOK AND CONCLUSION

The reciprocity of the relation between TermNet and INTER-CONCEPT is implicit in the definitions of their ultimate goals:

The ultimate aim of INTERCONCEPT is "to develop international understanding and co-operation in the social sciences; to support international comparative studies by providing the comparison of terms and concepts, and to ensure the conceptual preconditions for effective international communication in this field".

The function of Infoterm's TermNet on the other hand, is to ensure co-operation of terminology work on a world-wide scale, including all fields and all languages. Considering also that co-operation among many different organizations at a language or a subject level (or both) has already led to the development of networks, such as ISONET (with respect to data pertaining to standardization in general), EURONET (the European network under the auspices of the Commission of the European Communities designed to provide immediate access to scientific and technological information), etc. it becomes apparent that great strides have been taken already with a view towards a world-wide information network, as envisaged by Unesco's General Information Programme.

However, it has also been recognized that terminological compatibility is an essential prerequisite for the development of information systems, as exemplified by the conclusions and recommendations passed at the Third European Congress on Information Systems and Networks in Luxembourg [9].

"Overcoming the language barrier is a sine qua non for information systems and networks if we conceive their function as extending the purely parochial dimension of small user groups in very narrow subject fields..."

Let us hope that the establishment of TermNet with close collaboration between such sub-networks as INTERCONCEPT and others mentioned already, will aid significantly toward achieving this goal.

REFERENCES

[1] FELBER, H. COCTA International standardization of terminology. An Outline. Pittsburgh: University Center for international studies, 1976, 12 p., A4 (CISS Working paper no. 71, May 1976).

[2] Unesco. UNISIST. Study Report on the Feasibility of of a World Science Information System. Paris: Unesco, 1971, 161 p., 210 x 268, pp. 96-97.

[3] KROMMER-BENZ, M. Infoterm - International activities in the area of standardized terminology. In: Proceedings of the symposium on the international information network for standards - ISONET. Paris: Unesco, 1977, pp. 67-71, A4.

[4] FELBER, H. Study on the development of a network for terminology information and documentation. Vienna: Infoterm, 1978, 45 p., A4

[5] Unesco. Report on the meeting for the evaluation of the "Study of the development of a network for terminology information and documentation, prepared by Infoterm. Vienna: Infoterm, 1977, 12 p., A4.

[6] See particularly Chapter 6 of [4].

[7] FELBER, H. Developing an international network for conceptual analysis in the social sciences. The INTERCONCEPT project of Unesco. Paris: Unesco, 1977, 13 p. + 3 Annexes A4 (SS/77/CONF. 601/4, apr. 1977).

[8] RIGGS, F.; VAJDA, E.; VASARHELYI, P. "Interconcept". The establishment of an international information system of social science concepts. Paris: Unesco, 1976, 26 p., A4 (SS/77/CONF. 601/2 dec. 1976).

[9] COMMISSION OF THE EUROPEAN COMMUNITIES. Third European Congress on Information Systems and Networks. Overcoming the Language Barrier. Volume 1. München: Verlag Dokumentation 1977, 675 p., 160 x 240.

F.W. Riggs
University of Hawaii
Honululu, USA

6.6.8 PRELIMINARY SUGGESTION WITH REFERENCE TO INTERCONCEPT

Draft: For comments only, not for quotation

Introduction

Every record in the INTERCONCEPT terminology bank will be a citation, containing both quotations and references--in addition to other kinds of information. The words, "citation," "quotation," and "reference" are sometimes used as synonyms, and their meanings overlap in a confusing way. It will facilitate the work of INTERCONCEPT to use these words as technical terms. Broadly speaking, we can refer to any text copied from a source as a "quotation." The quotation in an INTERCONCEPT record includes both a term and its definition. Any indication of the source from which a quotation has been taken is its "reference." Typically, in an INTERCONCEPT record, this will take the form of a code number or tag, referring the user to an authority file that provides the full bibliographic information.

Although the word "citation" can be used for any record that contains both a quotation and a reference, we can also use it to refer to any part of a "citation." To avoid ambiguity, we will refer to any text that includes both quotation and reference as a "full citation," and we can then think of quotations and references, or fragments thereof, as "partial citations."

In order to indicate that something in a text is a quotation we ordinarily use double quotation marks ("..."), and sometimes single quotation marks ('...'). Other marks that can be found on any standard typewriter for English are: the question mark (?),the asterisk (*), parentheses ((...)), and underlining (...). It facilitates terminological and conceptual analysis to use these marks consistently whenever we use words relexively, i.e. to talk about or mention them. To speak of the word, 'development,' as distinct from talking about development, is to use the word reflexively. Sometimes linguists make this distinction by talking about "mention" vs. "use." To show that expressions are being used reflexively, in various ways, it is convenient to use all of the marks mentioned above. Since the word "use" is normally construed very broadly to include "mention" we might employ "proper use" less ambiguously to contrast with "mention" or "citation" as different kinds of use, or "reflexive use."

In part A of this note, the different types of citation will be discussed, and in part B the appropriate uses of each mark will be analyzed in such a way as to distinguish between its employment to indicate the reflexive use (or mention) of an expression and its employment in a context of proper use.

A. FORMS OF CITATION

Whatever a citation refers to is an "utterance." This is a very broad concept that includes both grammatical and logical types of utterance. Grammatical utterances include words, phrases, sentences, paragraphs, and complete texts. By contrast, logical utterances are classified as definitions, propositions, arguments, etc. Here we will focus on cited utterances, i.e. utterances which are mentioned rather than used (properly).

If an utterance that is mentioned includes both a quotation and a reference to its source, we can call it a "full citation." If it includes only part of a full

---

*The material in this note reflects the conventional practices of linguists and semanticists as reported in the recent and authoritative work by John Lyons, Semantics (Cambridge University Press, 1977) Vol. I, pp.5-13. In addition, some proposals are added to meet the special requirements of INTERCONCEPT.

citation, it is a "partial citation." However, if the partial citation is a quotation, then let us call it a "quotation," not a "citation." Similarly, if a partial citation is a reference, then let us call it a "reference," not a "citation." It will facilitate our work to avoid calling references and quotations "citations," because we can then use the word for a narrower concept, namely the abbreviated forms of quotations and references. When we speak of "partial citations," then, we will not mean a full quotation, or a full reference, but only an abbreviated form of either. To summarize:

(1) A full citation is an utterance that includes both a quotation and a reference

(2) A quotation is a cited utterance

(3) A reference is a bibliographic name of a source (ST: bibliographic reference)

(4) A partial citation is any part of a full citation not referred to as a quotation or reference

Note that in (3) the form, "ST" occurs as an abbreviation for "synonymous term." Since the word "reference" has a broader use that is relevant to our concerns, it is desirable to have a less ambiguous form that can be used for the same concept whenever it is not clear from the context that the narrower meaning of a term is intended.

1. Full Citation. It is not difficult to recognize a full citation whenever it occurs in a text, and the standard use of quotation marks and bibliographic conventions needs no further comment. In the INTERCONCEPT format, the status of the components of a full citation are clearly indicated by the program fields. In this format, quotations are distributed between field #04 for the term or definiendum; and #09 for the definition (definiens). The source is represented in field #08, either by a full bibliographic reference or a tag--i.e. a code number pointing to the full reference in the authority file.

2. Quotations. Any utterance copied from a source is a quotation, and is appropriately designated by quotation marks, following standard conventions. Little need be added except a reminder that, although technically a quotation is a partial citation, we shall reserve this term for other kinds of partial citations.

3. References. Conventions for the representation of bibliographic entries are well known and complex. They need not concern us here. In most scholarly texts, references are given in footnotes or in a list at the end of the text. A tag in the text itself serves as a "reference mark." Since we shall use the asterisk for a different purpose, the proposed convention is always to use reference numbers as reference marks. Another convention is to give abbreviated references in the text, enclosed by parentheses. The full text of references must then be given elsewhere, normally at the end of the text.

4. Partial Citations. Most of this note is concerned with the use of partial citations. Two kinds will be distinguished, those referring to quotations and those that refer to references. They will always be abbreviated or shortened forms, or numbers and other symbols pointing to quotations and references. Any partial citation to a reference will be called a "tag" and any partial citation to a quotation, a "lexical citation."

5. Tags. Consider the following utterances:

(5.1) David E. Apter, Some Conceptual Approaches to the Study of Modernization. (1968) p.330.

(5.2) (Apter, 1968. 330)

(5.1) is easily identified as a reference, and (5.2) is an abbreviated reference. Both come within the scope of the discussion in #3 above. We will find references of type (5.1) in footnotes or a list of references at the end of a text; and we will typically find abbreviated references, like (5.2) in the body of a text. By contrast, consider the following:

(6.1) (Apter)

(6.2) (5.1)

Both of the utterances in (6) are tag citations. In order to talk about them we need to define a few of the terms that we are using:

(7.1) A tag* is an expression used to refer to an individual item of reference, i.e. a referent.

(7.2) A bibliographic tag* is a tag referring to a source.

(7.3) A source* is a text from which information is taken.

(7.4) A text*is anything readable. Examples are: phrase, sentence, document, article, book, machine-readable file

(7.5) A tag citation* is a partial citation that uses a bibliographic tag

There are different reasons for and ways of referring to "Apter." If one wishes to talk about the man, one would use his name without any marks. But if one wishes to refer to a book he has written, using it as a source, then a reference is appropriate, as in (5). However, we may want to mention an idea contained in this source. For example, in the source referenced by (5.1) there is a definition of "development." We can precisely and easily identify this particular sense of the word "development" by using a tag after the word, e.g. "development (Apter)" or "development (5.1)". The intention here is not to direct the reader's attention to the source, but rather to specify which of several meanings of a word is intended. Ambiguity is overcome by calling attention, by means of a tag, to a reference in which the definition of the intended sense has been given. These, then, are "tag citations" in the forms given in (6). Another way of using tag citations is to use them properly. In the previous sentence, for example, "(6)" was used properly. If we wished to talk about "tag" in the special sense in which it is used here, we could refer to (7.1); this would be a proper use of "(7.1)" as a tag citation.

A useful distinction can be drawn between internal and external sources. From the point of view of any "given" text, other texts are "external." Normally a quotation included in a given text is derived from a source that is outside of that text. However, when a quotation is used in a text, it becomes a part of that text, and hence it becomes an internal source. In subsequent references, one may choose to refer to the quotation given in the text, rather than to its external source. A convenient way to do this without ambiguity is to number the quotation, i.e. to tag it, and then to use a tag citation to reference the quotation. In general, any utterance can be included and tagged in a given text, and then the tag can be used as a citation. If the citation is to an external source, its tag can be called an *external tag,* but if an utterance contained in the given text is referred to, then its tag is an *internal tag.* It is also convenient to use an author's name for external sources, e.g. (Apter), but to use numbers for internal tags, e.g. (8), which is a quotation from (5.1).

(8) "a continuous process of differentiation and increasing complexity"

A useful distinction can be made between "tags" and "names." Broadly speaking, any sign that signifies a concept, whether a general or individual concept, can be called a "name." More narrowly, if a general concept is named, we call it a "term," but if an individual object or entity is named, we can refer to its name as a tag*. If the tagged entity is a person, place, or organization, we typically call its tag a "proper name," and show this orthographically by capitalization, e.g. "Apter." Unless ambiguity is likely, it is normally convenient to limit the use of "name" to this sense of a "proper name." If this is accepted, then we can also limit the use of "tag" to the names of individuals that are not proper names. When the object named is a source, the name ofthe source is a reference, and the name of the reference is a bibliographic tag. The reason for distinguishing between the use of "partial citation" to refer to tags as distinct from references should be apparent. A tag names a reference, as a reference names a source. Of course, a tag can also name a quotation, which is also a citation from a source. When the tag names a reference, we call it a"bibliographic tag" . It is more awkward to name tags that cite quotations, due to the lack of an adjectival form of 'quotation,' but we could perhaps use, "quoting tag." Any tag that is used in a citation is a "tag citation," whether it is a bibliographic tag or a quoting tag.

Not all tags are bibliographic or quoting tags--some of them will be mentioned later in this note. However, tag citations are probably always either bibliographi or quoting tags. Normally it is unnecessary to specify, since the context will make the meaning clear. Thus (5.1) cites a bibliographic reference, and hence it is a bibliographic tag, whereas (8) is obviously a quoting tag.

As the foregoing examples have indicated, tag citations are enclosed in parentheses, a mark that clearly indicates their status. Other uses of parentheses which need to be distinguished from tag citations are discussed below, in part B.

6. Lexical Citations. Anything that relates to the words or vocabulary of a language is "lexical." Accordingly whenever we want to mention a word or expression we use a *lexical citation.*

(9) A lexical citation* is a partial citation that mentions a word or expression.

A basic distinction has to be made concerning the status of the expression to be cited. Normally we think in terms of a "words-to-meanings" paradigm, but sometimes the sequence is reversed and we think, instead, according to a "meanings-to words" paradigm.

(10) A words-to-meanings* paradigm considers the meanings of words in accordance with established usages.

(11) A meanings-to-words* paradigm assigns meanings to words in accordance with concept definitions.

(10) is the normal paradigm for all language sciences, including linguistics and most of its subject fields, such as semantics, lexicography, sociolinguistics, as well as relevant fields of philosophy such as logic, the philosophy ofscience, and the philosophy of language. By contrast, scientists, including social scientists and also language sciences when they are handling new concepts, rely heavily, though not always consciously, on (11).

On the basis of these two paradigms, we can distinguish fundamentally between two kinds of lexical citation. When (10) is invoked, we may speak of a lexicographical citation* and when (11) is the framework, of a terminological citation*. Let us take up the latter first.

7. Terminological Citations. We all know that words have been invented by the users of language, and it is scarcely news that science is responsible for the deliberate invention of many new words. A paradigm shift occurs whenever a new term is accepted and becomes normal usage. For the purposes of conceptual analysis, it is important to remember the early stage when new terms are being coined, even though most of the terms we know have, obviously, already been established. No doubt most of the terms used by scientists have been accepted already, but the

creative or innovative scientist, who is making discoveries, finds it necessary to define and use new concepts and, inevitably, then confronts the problem of what to call each such concept. One possibility is to coin a new word, and another is to use an old word but to give it a new meaning. Each strategy creates its own problems. My purpose here is only to stress that when a term is assigned to a new meaning, typically by a "stipulative definition," the paradigm that is relevant is (11), not (10). Unfortunately most of the discussions of stipulative definitions are framed in terms of the inappropriate paradigm, (10).

Using the convention proposed by Lyons and others, we can mark terms whose meanings are stipulated with an asterisk:

(12) Terminological citation*: mention of an expression that has been designated as a term for a concept.

Frequently an expression used as a term has a variety of meanings. We have to rely on the context to let us know which of the possible meanings is intended. Because of changing contexts it is convenient to be able to use more than one term for a concept. We have already used a number of terms that could be mentioned as examples, e.g. "tag" and "bibliographical tag;" "name" and "proper name;" "development" and "development (Apter)". Such synonymous terms* belong to a term family*.

(13.1)Term family*: a set of synonymous terms used to signify a single concept

(13.2) Synonymous term*: a member of a term family (also: ST)

In the words-to-meanings paradigm, synonyms are words with very similar meanings. A set of synonyms is a "synonymy." By contrast, if two different words are used to signify the same concept, the fact that they mean the same thing is stipulated. Since ordinary language words typically have a multiplicity of meanings, it is true that two words almost never have identically the same set of meanings.

The reasons for selecting more than one term to signify a concept arise from contextual considerations. The examples mentioned illustrate this point. In this note, so far, we have only used "reference" to mean a "bibliographical reference," although we pointed out the need for the synonymous term in (3). Broadly speaking a reference is a relationship between a referring expression and a thing referred to, i.e. the referent. However, the concept of a "bibliographic reference" is that of an entry that serves as a referring expression to refer to a source. In a given context of use, to use "reference" to mean a bibliographic referring expression is quite unambiguous, but in other contexts, where its meaning might be ambiguous, the more unequivocal term should be used. Accordingly we can distinguish between members of a term family that are shorter and easier to use, but possibly ambiguous out of context, and other terms in the same family that may be more cumbersome but have the advantage of being unambiguous out of context. This leads to the identification of two useful concepts:

(14.1) Synonymous term in context*: a term whose use is unambiguous in context but probably ambiguous out of context (also: STIC)

(14.2) Synonymous term out of context*: a term whose use is unambiguous out of context (also: STOC)

As a further illustration consider the word "development." We can imagine that in Apter's book, after having defined the meaning to which he has assigned this term, his readers will understand that when he mentions "development" he means (8). In an INTERCONCEPT context, however, the word "development" has been shown to have many meanings, one of which is Apter's. Accordingly the expression, "development (Apter)" is an unambiguous synonymous term for the same concept out of context. Accordingly, after (8) we might have written: "STIC: development; STOC: development (Apter)".

It is occasionally true that an ordinary language word has only one meaning. An example is "asterisk." It is possible to consider such a word, out of context, as a term. However, most words, because they have multiple meanings, can only be regarded as terms when they are used in contexts that make their intended meaning clear. To illustrate, consider that "star" is sometimes used synonymously with asterisk, but this word clearly has other and more usual meanings. Hence, although "star" may be a "synonym" for "asterisk," it is not a synonymous term. However, the expression, "star as a reference mark" could be a synonymous term for "asterisk."

Whenever we define a concept we have to use words in the definiens which may well be ambiguous. If one of the possible meanings of a word is intended, then it is necessary to make sure that the appropriate meaning can be determined, either by the context in which the word is used, or by a separate definition. To illustrate in (7.2) the word "source" was used in the definiens. However, the senses of "source" include the point of origin of a stream, a primary document, a generative force, and an initiator. To make certain that the intended meaning is clear, the definition in (7.3) was added to (7.2). Similarly, since "text" has several meanings, its intended meaning in (7.3) is given in (7.4). Any such word that is used in defining a concept is:

(15) An entailed term*: a term used in the definition of a concept

If the meaning of an entailed term* is unambiguous, then clearly no additional definition of the entailed term is required. We may refer to such terms as primitives* in the sense that no definitions are required to establish their meaning in the context of the definition. Normally it is easy to find out which of the senses of the word as given in a dictionary definition is appropriate.

Normally it is necessary to use the asterisk to mark a term only when its meaning is being stipulated by definition. However, if the word used as a technical term is unusual or difficult, it is quite appropriate to use the asterisk in subsequent applications just to remind the reader that it is, indeed, being used in a way that has been stipulated, and not for any of its other possible meanings. In the preceding paragraph, "entailed term*" was used with an asterisk after it had already beem mentioned in (15). However, when the term was repeated in the second line, the asterisk was omitted. Hopefully readers will be able to remember the meaning of "entailed term" without the reminder any more.

Since social scientists frequently (almost always, indeed!) use familiar words as their technical terms, the result is that many ordinary words become heavily overloaded. "Development" is an excellent example. It has become virtually impossible to remember which of the many possible meanings of "development" is intended, even when an author carefully defines the word as he/she intends to use it. If Apter had written "development*" rather often after he defined his term, I suspect his readers would have been helped to remember exactly how he wanted the word to be understood. Failure to adopt this convention means that much social science writing is indeed muddy and confused. The coding of abstracts for papers given at the Uppsala congress of sociologists showed that many authors were not only using "development" for different concepts but were, indeed, using it for several meanings within the context of a single abstract!

Sometimes an expression used as a term consists of more than one word. If it is not clear in context just which words belong to the expression, one can use an asterisk at the beginning as well as at the end of the expression, in the same way that quotation marks are used. For example, the paragraph preceding (8) (on p.3) uses *external tag* and *internal tag*. However, in (7.2) it seemed unnecessary to write *bibliographic tag* since the format showed clearly that the expression being defined contained both words.

8. Lexicographical Citations. The scientific study of words and their forms, meanings and uses is called "lexicology," and application of lexicology that leads to the preparation of dictionaries is called "lexicography." Accordingly, while it is appropriate to call a "terminological citation" a "lexical citation," we should limit the concept of "lexicographical citations" to those that might be based on the work of lexicographers. These always involve reporting on the use of words that have already become established. Accordingly we may say:

(16) A lexicographical citation*: a lexical citation that mentions a word or expression in accordance with established usage

Note that the underlying paradigm of (16) is (10), whereas the paradigm of (12) is (11). We can perhaps further clarify the difference between (12) and (16) by reference to a simple matrix, as follows:

(17) The Word/Term Matrix

	word 1	word 2	word 3	word 4	etc.
sense 1	a	b	c	d	
sense 2		e			
sense 3		f			
sense 4		g			

When we talk about concepts and their terms, using terminological citations, we have in mind situations like the one represented by (a, b, c, & d). A single concept, as specified by a definition, can be represented by several synonymous terms, constituting a term family. In the matrix, this is like "sense 1" that can be signified by word 1, word 2, word 3, and word 4.

By contrast, when we talk about words, we have in mind a situation represented by the letters (b, e, f, & g). A single word, e.g. 'word 2,' is said to have several senses--sense 1, sense 2, sense 3, and sense 4--as they are typically presented by lexicographers in ordinary dictionary definitions.

There is, of course, an overlap. Sense 1 can be designated by word 2 as well as by other words; and word 2 can signify sense 1, as well as other senses. To speak of word 2 as a term is to use it for only one of its possible meanings, e.g. sense 1; but to speak of word 2 as a word is to think of it in relation to its various possible meanings, notably (e, f, & g) as well as (b).

There are several ways that lexicographical information can be used in citations. In order to appreciate the important differences, let us take a look at ordinary dictionary definitions as though they were full citations with three significant parts. To illustrate, consider the following:

(18.1) de-vel-op-ment. n

(18.2) development, developmental, developmentally, developing

(18.3) development. 1: the act or process of developing; 2: the state of being developed

In each item under 18 something different is being said about "development." We can distinguish between these parts of an ordinary dictionary definition as follows: (18.1) tells us something about the form of a word, how it is spelled, which are its syllables, what part of speech it is, and how it is pronounced. (pronunciation was omitted from (18.1) due to the limitations of a typewriter!) (18.2) gives us several forms that have a common root and can be defined as a single word. (18.3) tells us about the different senses of the word. Accordingly we can distinguish between three types of lexicographical citation as follows:

(19.1) Form citation*: a lexicographical citation that mentions one of the forms of a word

(19.2) Lexemic citation*: a lexicographical citation that mentions a lexeme as a set of word-forms

(19.3) Semantic citation*: a lexicographical citation that mentions a lexeme in relation to its sense

Let us consider each of these types of lexicographical citation in turn.

9. Form Citations. If one is interested in morphology or plays Scrabble, one may be much interested in "word-forms." To say of development that it is an 11-letter word, that it has four syllables, that it is a noun, or that it is pleasant sounding, is to handle it as a word-form. To show that a word-form is being cited, it is a convention to underline it--or to put it in italics. It is rarely necessary to use form citations--i.e. the citation of word-forms--in conceptual and terminological analysis. However, it is important to understand the concept of a "word-form" and to be able to recognize what is going on whenever anyone mentions a word-form, i.e. makes a form citation. In the context of the INTERCONCEPT pilot project, it is clear that develop, development, developing, and developed are all different word-forms. Conceptual distinctions do hinge on differences between these forms. Thus develop is used to refer to a process, whereas developed signifies the outcome of this process. The meaning of a form used as an adjective often differs significantly from the meaning of the same form used as a noun. Moreover, knowing what word-forms are makes it easier to explain what a "lexeme" is.

10. Lexemic Citations. Normally when we talk about the meaning of a word we are not thinking about a single word form but about a set of forms such as we see in (18.2). Without becoming more technical, we can simply report that linguists use the word "lexeme" to refer to a cluster of word-forms that share a root-form. The form, develop, for example, is found in all the other word-forms included in (18.2). Whenever we want to mention such a cluster of words as a single entity, i.e. as a lexeme, but we do not want to refer to the meaning of the lexeme, we can make a "lexemic citation."

The standard way to mark a lexemic citation is by means of single quotation marks. Thus to refer to 'development' is to refer to all the word forms in (18.2)--and others as well, such as developmental, underdeveloped, overdeveloped, developmentally--but not in the context of meaning. We may want to identify the various word-forms that belong to the lexeme, and discuss their grammatical (paradigmatic and syntagmatic) functions. We may want to discuss their subjective or emotive "connotations," how people feel about them.

Obviously it is cumbersome to mention all the word-forms that belong to a lexeme when we want to cite it. Accordingly, it is conventional to choose one of the forms and use it as a *citation form*. Thus the citation form for (18.2) is 'development,' it being understood that the use of single quotation marks shows that not only development is being mentioned, but also all the other forms in the set.

Quite often we find two or more word-form combined in a sequence and used as though they constituted a single word. Lexicographers, indeed, treat such word combinations as entry words. Examples that come to mind are political science, social contract, city manager, social development, developing country, development planning. These are clearly word-forms, even though composed of two words. One way of referring to them is to call them "expressions." However, the word "expression" also carries a referential meaning, as we have already used it in the second paragraph between (13.2) and (14.1). To distinguish between the signification of an expression and its form, we might use "referential expression" in case of possible ambiguity for the former meaning, and *lexical expression* for the latter. The point to be made here is that word forms in a lexeme are often found in lexical expressions that need to be considered in the same context as members of the lexeme proper. In the pilot project we found it important to examine political development, economic development, and social development just as much as the various word-forms in 'development.' In order to be able to refer conveniently to the cluster of lexical expressions that are related to a lexeme, we might use the term, *expanded lexeme.* Accordingly we could understand a lexemic citation to involve mention of an expanded lexeme or of a lexeme by itself, i.e. a *lexeme proper.*

In the context of conceptual and terminological analysis it will frequently be necessary to mention lexemes (in proper or expanded forms) and so it is important to understand the concept and to use and recognize the convention of using single quotation marks.

11. Semantic Citations. By all odds the most frequent context of use for lexicographical citations, however, involves mentioning a lexeme or an expression in the context of its meaning. The convention to be used is to mark the word cited with double quotation marks. Moreover, accompanying words can be used to make the context of use doubly clear. For example, if we write, "the word, 'development,'" then it is apparent that a lexemic citation is intended, but if we say, "the concept of "development"", then clearly a semantic citation is intended.

When we move from the lexemic to the semantic level, we have to consider sets of expressions that refer to the same or similar referents. This brings us to the subject of synonymy, which was referred to after (13.2) Again, looking at dictionary definitions, we will sometimes find a group of synonyms listed at the end of an entry, followed by indications of the different senses or connotations of each. In the pilot project it was clearly important to examine not only the meanings of "development" but also related words such as "modernization," "progress," "evolution," and "change." The standard term for any set of synonyms is a "synonymy." Accordingly we can distinguish between a narrower and a broader type of semantic citation: in the former only members of a lexeme are cited, whereas in the latter, members of a synonymy are also included in the citation.

There seems to be no standard convention for distinguishing the different levels of semantic citation. However, it would seem to be useful in conceptual and terminological analysis to know what kind of cluster of word-forms is to be mentioned. The distinction could perhaps best be made by accompanying words,

(15.1) Semantic citation to a lexeme: to speak of the lexeme, "development"

(15.2) Semantic citation to an expanded lexeme: to speak of the expanded lexeme, "development"

(15.3) Semantic citation to a synonymy: to speak of the synonymy, "development"

12. Summary. To illustrate and summarize the various types of lexical citations, consider the following:

(16.1) development is derived from the French word, developper

(16.2) 'development' is a polysemantic word

(16.3) "development" means, among other things, the condition of being differentiated

(16.4) development* means a continuous process of increasing differentiation and complexity

All the utterances in (16) contain lexical citations. (16.4) contains a terminological citation based on a stipulative definition. Development* has to be understood as the form selected by an author to serve as the term for the specified meaning. The paradigm is a meanings-to-words perspective (11).

By contrast, the utterances in (16.1 - 16.3) contain lexicographical citations. (15.1) is a form citation; (15.2) a lexemic citation; and (15.3) a semantic citation. As noted above, conceptual and terminological analysis will have many occasions to use semantic and terminological citations, less frequent uses for lexemic citations, and rather infrequent uses for form citations. In the INTERCONCEPT context, the typical citations for target concepts will be terminological, whereas the typical citations for intermediate concepts will be semantic.

As a matter of concept clarification, it should be noted that many logicians use the term, "nominal definition" to refer to both descriptive (lexicographical) definitions and stipulative definitions. I believe it is a great mistake to think of terminological citations as though they were based on the words-to-meanings paradigm. Rather, they are based on the definitions of concepts, which implies instead a meanings-to-words paradigm. Accordingly the term, "nominal definition" should be avoided because it puts in one class two things which are so dissimilar that confusion of thought is bound to result. As for the term, "stipulative definition," it should be used with great caution because it directs attention to the arbitrary act of assigning a meaning to a term whereas it is far more significant to point to the creative aspect of such definitions in identifying new concepts. Of course, the new concepts may be worthless, but if so the judgment should be based on the empirical denotativeness of the concept and its theoretical relevance, but not on the form chosen to serve as its term (or terms).

A term that is often paired with "nominal definitions" is "real definitions," and a word should be added here on this expression. The form of a real definition is as follows:

(17) development means a continuous process of differentiation.

In (17) the word "development" is not being mentioned, it is being used properly. In this sense, the subject of the sentence is the referent (or denotation) of "development," rather than the word. In effect a proposition or an analytic statement has been offered. This type of "definition" goes back to Aristotle and Plato, and it has remained a perennial subject of philosophical discussion and controversy, up to the present day. What is being defined in a "real definition" is not a word but a thing. Predications capable of empirical verification are made in the expectation that they will identify the "essence" of the thing being defined. No doubt it is extremely important and useful to examine things carefully in the manner of "real definitions." However, as Robinson has shown, there are a good many different kinds of purposes and formulations that have been called "real definitions." (R. Robinson, Definitions. Oxford, 1950. p.149 ff) What is more to the point, I think, one cannot use a term to refer to anything until the meaning of the words used as a term have first been defined, either as technical terms by definition, or by context of usage. In this sense the ability to write a "real definition" depends on the prior knowledge of a lexical definition (whether terminological or lexicographical). If everyone would adopt the convention to use appropriate marks for lexical citations, it would be easy enough to tell whether words or things were being defined, since no marks would be used for the definition of things. However, since we cannot expect that everyone will use these marks, an alternative strategy would be to use some other term than "definition" for "real definitions," e.g. to call them propositions or analytic statements. The intent of this proposal is not to reduce the importance of "real definitions," but only to insist on the need to be clear concerning the status of words in use, properly, as contrasted with their status when being mentioned, in citations. The content of science is propositions and analytic statements, but the necessary tools are clearly defined concepts, and hence unambiguous technical terms.

. MARKING AMBIGUITIES?

It may be said that the proposed scheme for marking citations overlaps several ell established conventions governing the use of marks, and ambiguity could well esult unless additional marks are used. The answer appears to be that it is not ifficult to determine, by context, which of several possible uses of a mark is ntended. In any case, it is useful to consider the possibilities.

1. Underlining. The most usual meaning of italics or underlining is emphasis. o doubt this will continue to be true. If a word-form is to be mentioned, here is usually some contextual indication. Consider the following:

(18) development is a word-form.

No doubt ambiguity is possible. One might think that both development and ord-form were being emphasized. However, anyone who knows what a "word-form" s would immediately recognize that 'word-form' had been underlined for emphasis, nd development was the cited form. However, if the risk of ambiguity seemed too reat, an alternative formulation might be used, e.g.

(19) the word, development,is a word-form.

Since form citations will rarely be used in conceptual and terminological nalysis, we may safely assume that, unless contrary indications are clearly iven, underlining will usually indicate emphasis. Of course underlining is lso used, conventionally, for the titles of sources, in the context of a eference. Here the context makes the significance of underlining quite unambiguous.

2. Asterisks. Traditionally asterisks have been used as reference marks. or the most part this usage has already disappeared, and numbers are used instead s reference marks. Accordingly the use of asterisks for terminological citations hould be completely unambiguous. The follow-up use of asterisks as a reminder hat the terms marked have been defined by stipulation should, again, occasion o ambiguity.

3. Double Quotation Marks. Since both semantic citations and quotations are ignaled by the use of double quotation marks, will any ambiguity result? Probably ot, because a semantic citation is itself a kind of quotation. If we have quoted source that defined "development" to mean "the institutionalization of political rganizations and procedures," (Huntington, 1973. 139) then we are simply quoting he definiendum rather than the definiens when we give a semantic citation. In uch cases it makes no significant difference whether we think of the utterance in uotation marks as a "quotation" or a "semantic citation."

This is not to say that there are no differences between semantic citations and uotations. One ordinary difference is purely formal in the sense that semantic itations are usually quite short, whereas quotations tend to be longer. This andy distinction, however, may be regarded as rather trivial.

A more significant and substantive distinction can be noted, arising from the dentity of the reference in the two usages. When we speak of a reference, we ypically have in mind a single source. By contrast, semantic citations imply a ultiplicity of sources. Lexicographers make this clear when they give examples f contexts in which the words they define have been used--the Oxford English Dict-onary provides extensive references to illustrate contexts of use. However, one ould not think of any such reference as the source for a semantic citation. Thus, f the source for a quotation is a collectivity of users, then we are involved with semantic citation, whereas if a single source is referenced, then it is a uotation. To make this distinction more sharply, we could characterize any rdinary quotation, based on a single reference, as a "specific quotation," whereas semantic citation, based on an aggregation of users, might be called a *general uotation*.

If this distinction seems valid, then we can see how the double quotation mark is justified for semantic citations which are, in fact, "general quotations," whereas ordinary quotations are "specific quotations."

A third use of double quotation marks involves the citation of terms used in a thesaurus. Although one could think of these terms as specific citations, since they are taken from a single source, it seems more reasonable to consider them in the same class as general quotations since the controlled vocabulary of a thesaurus is designed to be used generally for retrieval and other purposes. In any event, whether thesaurus terms are thoughtof as general or specific quotations, they are surely to be considered in a context of their meanings and hence semantic citations, with the double quotation marks are assuredly appropriate.

4. Single Quotation Marks. Another possible source of confusion arises from the proposal to use single quotation marks not only for lexemic citations but also for index terms. As the analysis of the abstracts for papers given at the Uppsala Congress has shown, conceptual and terminological analysis can benefit from the systematic study of terms found in indexes. If we look at the Sociological Abstracts we will see, for example, such index terms as, "organization -s, -al, -ize, -izers, -ized, izing,' and 'sex, -es, -ism, -ist, -ists.' Quite clearly the intention here is to represent not a single word-form, but a complete lexeme, used as an index term.

However, it may be argued that index terms should be handled like terms in a thesaurus and mentioned in the form of semantic citations. Alternatively, it could be argued that the terms indexed were used in the text and so they are, in effect, quotations. In either case, should they not be cited with double quotation marks?

My opinion is that each entry in an index is, in miniature, a full citation. The index term is a quotation, and the page number a reference. To mention an index term, therefore, is different from mentioning a term in use, and is more like mentioning a lexeme. Indeed, to my mind there is no substantive difference between mentioning a lexeme and mentioning an index term. Hence the same mark can be used for both without danger of ambiguity. What is more, the convention of citing index terms with single quotation marks serves to distinguish this class of lexemic citations from the semantic citations of the same words when they are used properly in the main text that has been indexed. In my analysis of the abstracts contained in Sociological Abstracts I found it quite important to separate the citation of index terms from the citation of words used in the abstracts themselves. My conclusion, therefore, is that the use of single quotation marks for index terms as well as for lexemic citations does not create ambiguity and does serve a useful purpose.

5. Parentheses. The most common use of parentheses is to enclose supplementary or explanatory materials, such as the following:

(20) His interest in development (no doubt a confusing word) was great.

It seems unnecessary to take time to discuss this familiar usage. In case of ambiguity, of course, dashes could be used instead of the parentheses.

A much more important use of parentheses for anyone engaged in conceptual and terminological analysis is apposition. Frequently when the meaning of a word may be ambiguous, we use a synonym to remove ambiguity.

(21) His interest in development (institutionalization) was great.

In 21 the word "institutionalization" has been placed in apposition with "development" parenthetically to show which of several possible meanings of the word is intended. This concept, attributed to Huntington, was mentioned above in #B.3. Clearly the use of apposition can disambiguate many very polysemantic words.

To prevent possible confusion with the use of parentheses for tags, it might be thought that apposed expressions can be identified by commas. However, the use of commas is more common and confusing than the use of parentheses. Moreover, there is an analogy between the use of parenthetical apposition and tags that supports the use of the same marks for both. Suppose we are using the word "development" as a technical term, in the sense offered by Apter (5.1). Using tags we could write, "development (Apter)" or "development (8)". Alternatively, we could also write "development (differentiation)". The former involves use of tags, the latter of apposition. The effect, however, is identical.

Parentheses are also used for references, notably for abbreviated references. To continue the previous example, we could have written "development (Apter, 1968. 330)". The technical effect would be to direct the reader to a full reference at the end of the text which, in turn, would lead to the source. The substantive effect is identical with the two previous usages of tagging and apposition. The difference, and the reason for choosing one option in preference to another, is that the abbreviated reference presupposed ignorance of the source and the entailed quotation. The tagged citation presupposed knowledge of the source, and this permits a more succinct and convenient form. The use of apposition, again, presupposes familiarity with the intended sense of the word "development," but provides a basis for separating this sense from others which the word might also convey.

Another way of using parentheses is definitional. This usage is particularly common when introducing acronyms. Instead of using a formal definition, one may write:

(22) Division for the International Development of Social Sciences (IDS)

The user will readily understand that the acronym "IDS" is synonymous with the full proper name given in (22). The same format can be used to introduce a technical term. In place of (13.1) I could have written: "a set of synonymous terms (term family*)."

Reversing the sequence of the definitional use of parentheses, one can see that its import is also appositional. Taking my example from (7.3), we could have:

(23) source (a text from which information is taken)

In(23)the word "source" is used in a particular sense but, fearing the reader may not know the intended sense, a parenthetical apposition is used to clear up the possible ambiguity. Thus the definitional use, (22) is, functionally, the opposite of the appositional use, (23). In both instances parenthetical marks seem to be quite appropriate and not likely to be confusing.

To summarize, we have seen that utterances in the form of tags, acronyms, references, technical terms, and simplifications can be used, parenthetically, to serve the functions of apposition, definition, and reference. All these uses differ, structurally, from the use of tag citations. To talk about an utterance, such as the material tagged by (23), I can simply refer to (23) and this is a tag citation. The use of this device has been frequent in this note, and it has greatly simplified the presentation and discussion of many key points. To illustrate, we may write:

(24) (21) treats "development" as a process requisite.

All the examples of tags, used in parentheses, may suggest falsely that whenever numbers are used they should be parenthesized. However, it is clear that alphanumeric symbols can be used as terms in a sense distinguishable from tags. For example, we might define a concept as follows:

(25) Type A* means a class of concepts of "development" viewed as a process requisite.

In subsequent references, the term, "type A" would be used in the same manner as any word to which a technical meaning had been assigned. For example,

(26) Huntington's concept of "development" belongs to type A*.

As with other technical terms, one can not only use them, but also mention them. For example, one could refer to the concept of "type A," and also to type A as a word-form. If the term were used in an index, we might identify it as the index term, 'type A'.

6. Summary. It may be useful to summarize the foregoing discussion by means of a simple table, separating the use of citation marks from other uses of the same marks:

	CITATION MARKS	OTHER MARKINGS
UNDERLINING	form citations	emphasis
ASTERISKS	terminological citations	(reference marks?)
DOUBLE QUOTATION MARKS	semantic citations	quotations, thesaurus terms
SINGLE QUOTATION MARKS	lexemic citations	index terms
PARENTHESES	tag citations	apposition, reference, definition

The scheme appears to be unexceptionable except, perhaps, that one might prefeer to list index terms as a form of lexemic citation. However, I think there is a psychological advantage to calling the attention of users to the utility of marking index terms, explicitly, with single quotation marks. It could also be noted that we might have classed "quotations" as a form of "partial citation." However, we have also discussed the advantage of not referring to "quotations" as "partial citations," which leads me to list them in the "other markings" column.

C. CONCLUSION.

No attempt will be made here to summarize the argument advanced in this note. The presentation of the various forms of citation (full and partial, quotations and references, tags and lexical, terminological and lexicographic, form and lexemic and semantic) and also the various types of marks and their uses (underlining, asterisks, double and single quotation marks, and parentheses) has been schematic and sequential. It may strike some readers as unnecessarily rigid. My own opinion is that these conventions should be treated like crutches or scaffolding--to be discarded when they are no longer needed. Whenever the context of usage make a meaning unambiguously clear, it is unnecessary to use citation marks. If I say, "development has 11 letters," it is unnecessary to underline the word to show that a form citation is intended. Unfortunately, however, if we fail to observe these conventions at the beginning of our work we may suffer the fate of a victim trying to walk without his crutches. Serious misunderstandings and ambiguity are likely to result. To commence operations, therefore, in the field of conceptual and terminological analysis, and to launch INTERCONCEPT projects, it seems necessary to establish the guidelines and conventions that have been presented in this note and to recommend, if not insist, that they be used. Surely the adoption of these fairly simple conventions--at least, they will seem quite simply once they have been understood--will greatly enhance the precision and impact of our projects.

G. Rondeau
Université Laval
Québec, Canada

6.6.9 THE TERMINOLOGY BANK OF CANADA

I have been asked to discuss the basic considerations that were taken into account in establishing the Terminology Bank of Canada.

It may be useful first to outline briefly the particular circumstances in which the Terminology Bank of Canada - which I will refer to from now on as the "TBC" - was created.

About 35 years ago, when official translation services began to become more important in Canáda, a small group of terminologists was formed and given the job of helping translators in their search for what could be described as "the right French equivalent for an English word or expression", because, of course, most if not all translations were done from English to French. The job of the terminologists - though they were not so named until seven or eight years ago - was not an easy one, since French terminology was hard to come by in Canada at the same time and had to be extracted from foreign literature or invented, thus giving translators and their clients the impression that terminologists were cut off from reality. It is understandable that this small group which had built what I am inclined to consider as the first terminology bank in the world - a huge cylinder containing 125.000 records - should inevitably come to be looked upon as working in an ivory tower, and to be more and more ignored by translators.

In many departments, unofficial terminology lists were stored in computers and circulated. What's more, a number of translation sections or individual translators kept their own terminology files which had no connection with official terminological services.

I have given this background to illustrate the necessity of our first guiding principle: *That objectives for the short, medium and long terms must be clearly defined and must be based on an analysis of the needs of all potential users.*

This is what was done in 1974. First, we tried to identify all the potential users of terminology. I visited a few departments, the ones that were known to be the most active in producing terminology, and came to the conclusion that although the majority of terminology users were translators, there were other users with different needs. Even terminologists found it difficult at that time to accept the idea that there does not have to be a connection between terminology and translation: the term "monolingual terminology" itself was a revolutionary one.

We were then led to establish a useful distinction between *terminological problems* that may occur in the course of a translation and *translation problems* that a translator may face. The criterion used to differentiate these two types of problems is very simple: in a translation problem a translator has free choice of expressions, whereas in a terminological problem there is, theoretically, only one way to translate a term from one language into another.

This allowed us to formulate the first objective of the TBC:

1.1 To devote the terminology bank entirely to terminology, and consequently, to store terms with their related information and nothing else.

The formulation of this objective meant the elimination from the bank of all words and expressions belonging to common language and all translation "recipes" used by translators and usually found in their personal working files.

It is worth noting that two years later a new concept of the bank was developed - one that did not contradict the original one, but allowed a wider variety of data to be stored in a larger system. This was to be the next generation of the TBC and I may come back to it later during the question period.

Potential users of the terminology bank included the following four groups:

a) translators working toward or from languages other than English or French, namely Russian, Chinese, German, and Spanish;

b) writers of technical texts in the author's mother tongue or second official language. For example, in the Department of National Defence, many instruction manuals and brochures had to be written in French;

c) language teachers of civil servants. A large-scale program had been launched in 1968-69 with a view to providing second-language courses of all types, in both English and French, with the result that hundreds of teachers found themselves in the position of having to develop their own teaching materials. Unfortunately, for reasons too complex to explain here, the language teaching operations and the terminology development programs were assigned to two different departments, the Public Service Commission and the Secretary of State Department. As a result, there was insufficient coordination and no LSP programs were set up;

d) readers of technical texts in French or English, regardless of their first or second language.

This led to the formulation of the second objective of the TBC:

1.2 To respond to the needs of all users in the Government of Canada, first in the two official languages and, as soon as possible, in other languages.

This second objective had numerous consequences, the most important of which was a complete change in approach to the "terminology record". Up until then, a record had been understood to contain one concept expressed in two languages (as illustrated below), in accordance with the definition agreed upon earlier at the Baie St-Paul symposium[2]:

TERM (in E) + Related information	
	Basic components: one concept and two terms, each in a different language.
EQUIVALENT (in F) + Related information	

But the second objective conflicted with this approach for two reasons:

a) restriction of number of languages to two; and
b) exclusion of monolingual terminology.

Consequently, the terminology record is now conceived as a single block of information containing one concept and its designation in one language. In other words, what was formerly considered a half-record is now regarded as a full record.

This will allow storage of monoligual terminology as well as multilingual without pre-determining the number of languages in which a concept is to be expressed in the bank.

However, I must note here that this new approach although widely accepted, has not yet become so much a part of terminologists' thinking that it has affected their vocabulary.

Many of them still use the terms "entry" and "equivalent", revealing a translator's attitude toward terminology.

Before turning to our second guiding principle, I would like to point out that when we were establishing the objectives of the TBC, we took into account user's needs as seen by users. To identify these, we destributed two questionnaires, one to various departments and one designed specifically for translators. The first one did not elicit much response, but the second was completed by a large percentage of translators. Briefly, most translators wanted a system that would:

a) give fast answers, whether affirmative or negative;
b) supply useful information that would help the user find a solution when the system could not supply the required term;
c) be used only when ordinary (non-computerized) tools failed to produce an answer;
d) be simple to operate and hence avoid burdening the user with added work.

This survey confirmed guiding principle No. 2: 2.1 *That documentation is an essential component of terminological services* and: *2.2 That documentation and terminologies should be classified according to the same system for optimum efficiency.*

One thing that must be said about the TBC is that because it was established fairly recently, its designers were able to benefit from the experience of others. We went through all available literature on existing banks, particularly those of Luxembourg and Quebec. With reference to documentation two points became clear:
a) terminological services cannot be offered without a proper documentation system;
b) the classification of documentation is a major problem that should be solved before any bank is put into operation.

On this second point, I must say that no satisfactory solution to the problem has yet been found. The working group known as "Mission terminologie" made a careful study of all existing systems and decided to reject them on various grounds. In 1975 they proposed a three-level system with 80 categories at the first level - later reduced to 50 or so. This system is still in the developmental stages.

In my view, a final solution will be found only if and when a universal classification of concepts is achieved, which would allow concepts to be defined by means of a set of relationships, thereby making it possible to classify texts or terms according to concepts.

Let us turn to the third guiding principle: *3.1 That all users of the TBC should contribute to the bank in so far as possible; and 3.2 That all terminological information, even if incomplete, is worth recording if it is of good quality.*

We thought, and we still think, that users of a terminology bank - terminologists, translators, language teachers, and so forth - can provide useful terminological information, even if it is incomplete. The problem is to convince them that, in the long run, the investment of a little time and effort will pay off. Of course, we do not expect them to be the sole source of terminological data; as I will explain later, they represent only one of many sources, and are by no means, the major one. But there is a psychological benefit to be derived from contributing to the bank - a feeling of belonging to a network. This principle has been applied in the case of our first users, translators. Here is how it works:

A number of terminologists have been specially trained for translation sections and have the twofold task of (a) helping translators with their terminological problems, either by querying the TBC or by contacting specialists, or by looking through documentation, and so on; and (b) collecting terminological information produced by translators in the course of their work. Translators are asked to write this information on specially designed "mini-records" that they can fill out in a minimum of time by limiting themselves to the essentials.

It is then the job of the "field terminologists" to complete these records and transcribe the information onto the official Termium II record before transmitting it to the bank.

This system has been in operation for only a few months, so I have no statistics here on which to base an evaluation of its efficiency, but it is being monitored carefully.

Personally, I feel confident that useful terminological data that would otherwise be lost or at best kept in the personal files of the translators will, in this way, be made freely available. (Of course, careful checking for quality is done before terms originating from users are entered into the bank.)

This principle was again applied in early 1975 when, with the agreement of those concerned, all the personal terminological files of translators were photocopied. This was done at night, in order not to disturb the translators' work. One million eight hundred thousand terms were collected in total. They were then typed for optical reading and fed into the computer.

Of course, many of the terms were repetitious, and a large number of entries had to be deleted for various reasons; this was done through two successive operations known as "pré-épuration" and "épuration" (the latter is still continuing), leaving about eight hundred thousand terms in the working file.

Let us now consider our fourth guiding principle: *4.1 That depending on its objectives, a terminology bank either must be normative and assign relative values to the terms it contains or must simply record usage, without making any value judgement*. There can be no mixing of the two.

The TBC adopted the attitude that it is more the job of lexicologists and lexicographers to simply record usage. We felt that a terminology bank had a social role to play in that it would be called upon to disseminate correct forms and terms or identify substandard elements that were, for a variety of reasons, included in the data base.

The second part of our fourth guiding principle read as follows: *4.2 That terminological standardization should be done by specialists in the field with the help of terminologists, and not vice versa.*

This, of course, is now a widely accepted fact and no terminologist today would attempt to do his job "en vase clos". But in the Translation Bureau three years ago the principle had to be reaffirmed because of a tendency among terminologists to consider themselves as the only ones in a position to attest to the validity of a term.

Our fifth guiding principle concerned the personnel of the TBC: *5.1 That the distribution of tasks within a terminology bank should be so organized as to reflect two types of terminologists: terminologists' assistants, or "terminographs", as Natanson and Rey have suggested, who would perform routine tasks; and terminologists, highly trained professionals who would be expected to make decisions when necessary. 5.2 That the TBC should provide in-house training, at least until such time as relevant programs are available in educational institutions.*

The application of 5.1 did not prove easy. In fact it has been the subject of lengthy discussion on numerous occasions, most notably by the last international symposium on terminology at Lac Delage[3], and is still under study at the TBC, although a number of persons have been hired, not as terminologists, but to perform jobs, such as rewriting, in the proper format, terminological data supplied on non-conventional forms, reviewing or gathering documentation on certain subject fields, checking the content and layout of records, and so on.

The application of 5.2 is under way. A training program complete with textbooks, readings and exercises is being prepared. The *Guide de Rédaction de la fiche Termium II/Guide for Completing the Termium II Record* is now in use.

Let us now turn to our sixth guiding principle: *That the software for the terminology bank should be designed specifically for it or should correspond exactly to its objectives, and should not be subject to the availability of commercially developed systems.*

To apply this principle it was necessary to take into account economic conditions, among other things. The following seemed to be the best way to proceed:

1. Definition of objectives
2. Detailed analysis of needs (present and future)
3. Comparison of existing systems in light of points 1 and 2
4. Feasibility and cost studies re adaptation of most suitable existing systems
5. Feasibility and cost studies re development of a new system
6. Comparison of findings in points 4 and 5
7. Decision

We started to go through these steps and ended up stopping at No. 4 because both the software and the personnel of the Banque de terminologie de l'Université de Montréal were available and the modifications required to adapt their system for our purposes, i.e. make it operational, did not appear to be too costly.

In short, the system that was in existence in Montreal in 1975 is currently referred to as Termium I; in 1976-1977 Termium II[4] was designed and it is now in the process of being implemented. Preliminary studies began a few months ago on an eventual Termium III system.

Our seventh guiding principle was: *7.1 That terminology banks should be equipped with control mechanisms that will allow constant and up-to-the-minute evaluations of different factors such as the ratios for cost/response time, query/silence and query/noise, statistics in the frequency of consultation for each entry, and so on; and 7.2 That it must be easy to update the bank.*

It was felt that it should be possible to perform these evaluations automatically, and that this would be very useful for improving the efficiency of the bank.

Today it is generally recognized that no single terminology bank can deal with all existing terminologies and must therefore belong to a network. Thus guiding principle No. 8 was: *That terminology banks should be designed for intercommunication with ohter terminology banks.*

There appears to be a conflict between principles No. 1 and 8 which may hamper the establishment of an international network. The situation could, I think, be summarized as follows:

1. Exchange of terminological data is possible on an individual basis, provided the receiver is prepared to invest in the computer specialist's time. Examples are exchanges between Montreal, Luxembourg and Quebec.

2. No terminology bank has, to my knowledge, undertaken any work to modify its format so that terminological data could be transmitted directly to another bank. This, however, would seem to me to be a prerequisite to the establishment of any network.

3. The trend, witness the Federal Republic of Germany MATER system and ISO/TC37/WG4 studies, is not to exchange terminological data but rather to design interface programs that enable a terminology bank to read the computer tapes of other banks. This suggests that the role of a terminological network would be to publish information concerning the location(s) of various terminologies.

NOTES
=====

1. Paper presented at the opening session of the International Experts Meeting on Terminology Banks. Copenhagen. December 1-2, 1977.

2. Baie St-Paul (Québec), October 1-3, 1972

3. Lac Delage (Québec). October 5-8, 1975

4. Some differences between Termium I and Termium II

Termium I	Termium II
Pre-set field length	Variable field length
Language identified by location of term	Language identified by code
One file containing only complete records	Three files
Limited space on the "input" record	Unlimited space

W.H.U. Schewe
Deutsches Institut für Normung (DIN)
Berlin, FRG

6.6.10 RECENT EXPERIENCES WITH THE ESTABLISHMENT OF A TERMINOLOGICAL DATA BANK WITHIN DIN

1 The importance of language for technology

As a result of the ever-increasing rate of progress in science and technnology in the 20th century, man is faced with certain problems which, as a user of language, he has to solve. Rapid progress in conjunction with international interdependence in all fields of science and technology involves such a large number of new features (devices, apparatus, machines, methods, scientific relationships etc.) that choosing and coining the specialist terms required for all these objects and concepts have become much more important than was previously imagined for communication within a given language area and even more so at international level.

The 'homo faber' of our century demands that language, his most important means of communication, should fulfil certain requirements to enable or at least help him to overcome technical problems. At the same time the flood of technical innovations increases the demand for new words to such an extent that we are now even heaving about the Technologist's "word shortage", and indeed we may agree with some linguists, that never has any field of human activity demanded such much of language as technology.

The main demand imposed on specialized technical languages externally, i.e. by their users, is the demand for accurate and unambiguous communication with at the same time maximum economy of language. This demand has necerialy led to obligatory designations in the various technical fields - i.e. by the means of standardization of terminology.

2 The importance of EDP for terminological work

Whereas terminological activities until now mostly have been performed manually, more and more national or worldwide new technologies are introduced to allow handling and evaluation of language. Particularly electronic data processing has become an important tool in all fields of terminological activities. The use of such equipment is not restricted only to the collection of terminological data but spreads over to storage, retrieval and other kinds of evaluation. Thus in several parts of the world terminological data banks have been built up or are in the process of establishment. "Terminological data bank" is to be understood in the following way: a computer-aided collection of expressions of at least one natural language which is usable for multiple purposes like retrival of specific terms, selection and publications of special dectionairies, translation etc.

Fields of activity in which terminological data banks come into use are particularly

- terminological work and standardization
- information and documentation (thesauri, classifications)
- dictionary preparation, printing, and publication (of all kinds)
- translation
- linguistic research

The compilations of terminological data in these fields can - in most cases - be used in other areas of application as well, e.g. special dictionaries are a very important source for thesauries construction. This implies that, if data of this kind are available in machine - readable form the exchange of these data between the different agencies could be a tool to improve the work of the respective agencies (by getting larger data bases for their work) and to save time and money. Such an exchange requires a standardized format, otherwise a lot of work has to be invested to make data formated for one purpose usuable for another purpose.

To facilitate exchange of terminological data on an international level is the aim of the work in progress carried out by ISO/TC 37/WG 4 dealing with the "Magnetic Tape Exchange Format for Terminological(Lexicographical Records (MATER)" - ISO Draft Proposal 6156.

It is intended that MATER should be able to handle the following kind of data:

- monolingual dictionaries and encyclopaedies
- multilingual dictionaries and encyclopaedies
- multilingual glossaries
- classified mono- or multilingual vocabularies
- mono- or multilingual thesauri.

The set of categories to handle all these data is very extensive in MATER. If MATER will be used in the form specified in the Draft Proposal 6156 it will have at least two advantages:

- The emitter of an exchange tape will know exactly, what structure and which categories he has to write on the magnetic tape.
- The receiver of tapes from different sources will need only one programme to read the tapes; it may also happen that he will have to do only slight modifications and must not do the entire programming work each time when a input tape will arrive from a new source.

3 Establishment of a terminological data bank within DIN

Having received a financial grant from the "Bundesministerium für Forschung und Technologie" (Federal Ministry for Research and Technology), DIN has started the establishment of the terminological data bank "TERM" taking into consideration the proposals of ISO DP 6156 "MATER". In close cooperation with SIEMENS AG in München (System "TEAM") the following set of categories has been elaborated which is the basis for the storing of terminological data extracted from the German Standards and Draft Standards:

TERM*	(ISO 6156)**	
300	(100)	Main entry
330	(130)	Abbreviation
602	(502)	Synonym
604	(504)	Quasi-synonym
700	(520)	Broader term (superordinated term)
790	(560)	Antonym
402	(402)	Definition
432	(432)	Example
434	(434)	Footnote
404	(404)	Explanation
440	(440)	Restriction
710	(530)	Narrower term (subordinated term)
300E	(100E)	English term
402E	(402E)	English definition
404E	(404E)	English explanation
432E	(432E)	English example
434E	(434E)	English footnote
700E	(520E)	Broader term for 300E
300F	(100F)	French term
300S	(100S)	Spanish term
300R	(100R)	Russian term
300I	(100I)	Italian term

---

*) Categories of TERM format (in accordance with Draft German Standard DIN 2341 MATER; May 1977)

**) Categories according to ISO draft proposal 6156 Magnetic tape exchange format for terminological/lexicographical records (MATER); July 1978.

115	(815)	Outline description of source (Number of standard, part, status (standard, draft standard or preliminary standard), edition)
120	(820)	Location in source (page, paragraph)
210	(910)	Agency responsible (standards committee responsible for the standard in question (coded); standards committee which is co-author of the standard)

A pilot project dealing with the terminology of information and documentation contained in the German Standards and Draft Standards has been finished. It has bevome obvious that most of the categories mentioned above are very useful for the description of terminological data.

DIN ist now endeavoured to establish a terminological data file which will store terminological data from all technical fields of standardization. This will be a great task for it has to be mentioned that up to now we have looked through ca. 800 German Standards and Draft Standards - from a total of about 17 500 German Standards and Draft Standards - which appeared since February 1978 and we have found there ca. 10 000 terminological determinations - that means terms and definitions.

We have also developed a bibliographical data sheet which contains the following information:

TERM	(ISO 6156)	
115	(815)	Outline description of source: Number of standard, part, status (standard, draft standard, or preliminary standard), edition

200	(900)	Copyright: DINP
940	(710)	Notation: UDC
116		German title of the standard
116E		English title of the standard
116F		French title of the standard
210	(910)	Agency responsible: standards committee responsible for the standard - coded according to DIN data banks code - full name of the committee
215.		standards committee which is co-author of the standard - coded - full name
A01		Indication whether the standard - contains definitions or terminological explanations (= TE) - contains vocabulary exclusively (= TD) - contains only terms without definitions, with English, French etc. equivalents (= TB).

It can be hoped that the above-mentioned bibliographical data will give some additional help for the searching of standardized terminology.

4 The relevance of a terminological data bank

It has been pointed out in a report of Mr. Felber on terminological data banks that these data banks will be a very useful tool for the "unification of concepts and for checking terminologies in order to avoid discrepancies as regards definition of the same term."

In addition we would like to summarize the following use of a data bank:

- help for terminologists within or outside a standardization committee (most complicated needs)
- help for translators (e.g. equivalency of terms)
- help for documentalists (e.g. elaboration of a thesaurus)
- help for technicians (e.g. unterstanding of the technical terminology)
- help for teachers, students, journalists (e.g. study of neologismes)

As we have pointed out the terminologists within or outside a standardization committee will have the most differentiated needs when working with a terminological data bank. There is up to now no detailed experience with the requirements which will be claimed by the terminologists but attention could be drawn to the fact that terminologists are generally using all information offered to them by means of electronic data equipment. The presentation and editing of specialized technical vocabularies could only be one aim of their efforts.

Technical languages come into being as a result of the need to be able to describe with the necessary accuracy what is involved in a given specialized field. Their structure varies in the same way as the technical facts they have to describe. The scientist and technologist uses the term or creates new terms in order to express and communicate to others conceptual associations more clearly and accurately than is usual in common language. But the main concern is always that new terms should not be formed arbitrarily but according to definite principles which ensure that the terminology is clear, unambiguous and uniform. The standardization of terminology according to the rules contained in the standards developed by ISO/TC 37 is therefore an essential part of material standardization, indeed often a prerequisite for it. The use of a terminological data bank will draw attention especially to this point. And it is for that reason that electronic data equipment can be regarded as a most basic tool for every terminological work.

Secretary of State
Terminology Directorate
Ottawa, Canada

6.6.11 THE CANADIAN GOVERNMENT TERMINOLOGY BANK

An overview (Revised March 1978)

Mandate

In November 1974, in view of the need to co-ordinate and rationalize terminological work in the Public Service, the Cabinet (in decision No. 569-74-RD) gave the Translation Bureau the responsibility of checking and standardizing French and English terminology in the federal Public Service, Parliament and its agencies. This confirmed and clarified an objective assigned to the Translation Bureau earlier that year by Treasury Board, namely, "in co-operation with specialized institutions in Canada and abroad, to organize and promote terminological research and set up a bank of terms and equivalents in anticipation of the needs of Parliament, the government and its agencies, to increase the efficiency of translation in all fields, and specifically in those of science and technology" and "to co-ordinate the work of terminological standardization in government organizations, co-operating in this task with all linguistic research and standardization centres in Canada and abroad". The mandate given to the Translation Bureau by Cabinet and Treasury Board involved setting up new structures to check and standardize the terminology used throughout the Public Service, and also devising methods of storing the terminology and redistributing it throughout the Public Service.

The federal government's translation services are still the largest group of users of terminology. As a result, it was reasonable to use them as a guide in the preparation of a study which found that a terminology bank would be economically profitable, in terms of the increased productivity alone. Other, non-monetary considerations also argued in favour of a centralized, computerized terminology service: a) the need to provide the staff of units working in French with reliable terminology; b) the need to standardize the quality of the manuals used as work instruments in the Public Service; and c) the need to standardize the terminology of the acts and regulations, which the Department of Justice had just undertaken to revise. Another significant factor was that some departments already had terminoloy "minibanks" of varying degrees of refinement, and this greatly increased the chance of needless expense, duplication of efforts and lack of standardization.

Acquisition of the bank

After five years of research and development, the University of Montreal had set up a terminology bank based on a specially designed software. At the end of 1975, the University decided that it was no longer its function to continue developing a system that had become operational, and offered to turn its terminology bank over to the federal government.

A study by a special task force found that this bank, which was designed first and foremost to meet the needs of translators, was the best suited to the Translation Bureau's needs. As a result, the Secretary of State Department accepted the University's offer and took over its bank. Some adaptations and changes were made in the software, and then in January 1977, the system and its contents were transferred from the University of Montreal's computer to that of the Department of Energy, Mines and Resources in Ottawa.

The changes in the software were designed to adapt it to new users and new needs, because the users will no longer be just translators, but all employees of the federal government, both translators and others who are called upon to write or adapt texts, who are involved in language training programs, or who administer acts and regulations both inside and outside the Public Service. Consequently, the bank must become universal and contain not only scientific and technical terminology, but also common terms that do not present any difficulties (for standardization or teaching purposes), as well as highly specialized terminology such as initials and abbreviations used in the Canadian Forces, and terms used internally by its various clients.

The bank today

From the outset, the Translation Bureau had a huge number of terminology records in its various sectors, and these records dealt with all the subjects in which a country the size of Canada might have an interest. For the most part, the records were prepared by translators who had worked in the Bureau since its creation in 1934 and even earlier.

Unfortunately, terminology in those days was not what it is now, and the records were not prepared according to a specific method. As a result, they assumed a variety of forms. In a great many cases, they simply consisted of a term in the source language and its equivalent in the target language, occasionally with an indication of the source and author. Consequently, the quality of these records varied greatly, and they contained inaccurate, insconsistent and contradictory information. Most importantly, there were a great many duplications. Despite all their shortcomings, these records constituted an initial transitional file that at least gave us an idea of the work to be done in the area of terminology and hastened the production of an acceptable file. On this basis, it was decided to load this transitional file after the changes had been made in the University of Montreal's software. The records already in the memory at the University of Montreal, plus some 150,000 records on various magnetic mediums in the Public Service, made a total of 1.7 million records for our first transitional file.

In order to store this wealth of information, we first had to standardize the records and classify the information they contained into the following eight basic fields:

- source language entry term;
- target language entry term;
- source language code;

- target language code;
- subject field;
- source of the record;
- originator of the record;
- "catch-all" field (for any other information that might appear on the record).

All these fields were compatible with the medium offered by the University of Montreal's bank.

Of the expected 1.7 million records, 1,233,000 have been loaded into the memory. As a result of the ensuing systematic preliminary screening and screening, the number of records in the bank was reduced to 800,000, all of which are accessible on-line almost instantaneously from about sixty terminals.

Termium II

Since January 1977, records have been prepared according to the Termium II software. These records are all single-concept records: users will proceed from one term to another. This will make it possible to group related terms (synonyms, abbreviations, spelling variants, and so on). All new entries related to concepts already in the bank will be additions to the existing concept record rather than new records.

As more and more of the transitional file is transferred to Termium II, and hence to concept records, we will gradually move away from the lexicographical approach and, within a few years, we will have a genuine terminological file.

If we had to describe the Termium II system in simple terms, we could say that it is the common terminological memory of all users. For the present, the users are the 1,200 translators in the Bureau and a few hundred other public employees. In the medium term, users will be from the entire federal Public Service and an unlimited number of other "clients". Terminology bank clients will not only be users, they will also be suppliers, and they will be expected to contribute data from their own stock of knowledge to the common pool. The data stored in the bank are not always perfect; in fact we wish to add day by day the data submitted by our users. We intend to load incomplete solutions and even unsolved problems. Since a large percentage of our users are specialists, we are entitled to expect them to help us complete and correct our files. The information we give users in response to their questions, even if it is incomplete, should produce a chain feedback reaction, which will enable us to complete and improve our file quite quickly.

We worked from the premise that even incomplete records are always of some use, because they contain at least a few scraps of valuable information. We are able to put incomplete information into the bank (something that could not be done with a dictionary), provided we establish a mechanism for completing and correcting it bit by bit, as more data is received. If we can get our users to let us know when they find our terminology inadequate, or to provide us with the solutions they find in a source other than the bank, the quality of our bank will be directly proportional to the number of users. This approach allows us to put suspect data into the bank, provided we include the necessary precautions, so that they are not presented as reliable information.

Screening

The screening procedures described in the overview of January 1977 were modified as operations progressed. Through systematic or random preliminary screening and screening, 400,000 records which were no longer useful have been eliminated from the Terminology Bank.

Today, screening is not conducted on a systematic basis, except in cases where an entry term occurs at least twenty-five times. In order to reduce the mass of information and thereby increase the usefulness and relevance of the bank, terminologists systematically examine the cases in which the same entry term is found in the bank over twenty-five times. They separate the records according to concept and make a new Termium II record corresponding to each of the concepts concerned. This may lead simply to a reorganization of the information which is already available in the bank or to new terminological research. In all cases, the number of new single-concept records is lower than the original number of records for the term. The original records are then eliminated from the bank and replaced by new ones.

Generally, screening takes place at the same time as other research work. The terminologists are required to carry out a systematic search of the Terminology Bank before undertaking research and, once they have entered the results of their research on a single-concept Termium II record, to request the elimination of multi-concept records which have been made superfluous by the new record.

The bank's structure

A diagram of the Terminology Bank's structure is provided in Appendix I.

The master file contains the terminological records which meet the basic terminological criteria and which may therefore be exchanged with other organizations and other banks in Canada and abroad. This means that they may readily be distributed to any category of user. The information they contain is accurate and complete and does not duplicate other data in the single-concept sub-files.

The transit file contains records in which all the terminological information given is accurate, although some doubt remains in the minds of the originators as to the precise equivalence of the terms in both languages.

In short, these are records which require additional details before they can be considered perfect. It may not always be possible to distribute them freely.

The working file contains records which were found to be single-concept records through prior screening of the bank, although there is serious doubt as to the accuracy of the information in them. These records are retained in the bank as "unsubstantiated information" for which the originator can give no guarantee of their terminological value or the equivalent suggested.

The titles and official designations file, by its very definition, is completely single-concept in nature. It contains only proper names and entries similar to proper names and includes titles of books, films, names of organizations, job titles and so on.

The Documentation Directorate, which has a general reference unit, already has a good proper names file, which will be used as a basis for the computerized file of titles and official designations. It will use the same format as that used for terminology, thus simplifying the input operation considerably.

The automatic dictionary is used to store simple equivalents which do not require terminological research. It is also used to load the terms and their equivalents extracted from bilingual publications of official organizations so that translators may consult them more quickly.

Finally, the research file contains all the other records which are considered useful for research purposes. These may be small pieces of information or complete records from other organizations working with different methods. They may even be taken from books which cannot be entered in the single-concept file. Few restraints are imposed with respect to this file, and it can be loaded with little preparation. Access to this file is obviously more limited.

<u>The terminology network</u>

A large terminology research service, currently 120 terminologists and support staff, will back up the computerized processing system. The work method used by terminologists is in accordance with guidelines established jointly with the Quebec Office de la langue française and private industry. These terminologists, together with the terminologists of the bank's clients outside the Bureau, form a terminology network, each component of which will have specialized knowledge of its client's subject area. All of them have direct access to the bank through terminals and can (and should) pass on specialized terminology problems raised by their clients to the appropriate component of the network. In this way, terminologists will always work in their area of specialization, and since they will all use the same research method, the product will always be of the same quality. The quality will be that much better if the terminologist has access to the specialized documentation of the section in which he works, and - another important factor - to specialists in the field in question.

Thanks to the flexible structure of the files, as we saw above, each terminologist will be able to load, with ease and for the benefit of all, any information which he has found, even before completing his research if necessary.

Our terminals are linked to the computer by telephone lines. It may become possible to give other terminals access to the bank, even on an occasional basis, which would be particularly interesting for services that already have terminals for other purposes and might sometimes want to consult the Terminology Bank.

Conversely, there is no reason why our terminals could not be used to consult other data banks. In fact, our Documentation Directorate is now looking into the possibility of giving four branch libraries access to documentary data banks through the Terminology Bank terminals. This would automatically ensure better service to our clients, since it would enable them to obtain both terminological and documentary information from one terminal.

Clients

At present some 70 terminals are connected to the Terminology Bank, most of them within the Translation Bureau. Eighty-nine per cent of federal government translators have access to terminals in the buildings in which they work, and 81% have terminals on the floors on which they work. The monthly number of consultations is 37,825, at an average cost of $0.55. The response rate is about 45% per cent. We hope to be able to increase the number of consultations and improve the response rate while lowering the unit cost through bulk loading of data from the National Defence Department and through systematic utilization of existing standards, glossaries and vocabulary lists.

The bank also has some clients outside the Secretary of State Department. Each of these clients has been required to employ terminologists who are capable of contributing to the terminology stock and make rational use of the stock, following training by our own services when necessary.

The network of terminals

There are at present 62 terminals in the network, 59 of which belong to the Translation Bureau and three to other clients. They are located in Ottawa, Montreal, Quebec City, Saint-Jean (Quebec), Moncton, Toronto and Winnipeg. In a few weeks' time, the Terminology Bank will acquire a new client. One of its translators has already received training in terminology from our services. In 1977, we acquired two high-speed terminals (one in Ottawa and the other in Montreal), which make it possible to produce vocabulary lists and other lengthy printouts. These high-speed terminals are able to print 900 lines a minute, compared with the 30 characters a second printed by the low-speed terminals in the network.

Last fall, a study group was set up to establish a medium-term policy on the expansion of the Terminology Bank in the light of the present operation of the network and the technical and financial constraints governing such development. A study was considered justified because of the increasing number of requests for terminals from various sources (Translation Bureau, federal departments and agencies, provincial governments, national and international organizations, private industry and universities) and the resulting need to ensure the rational expansion of the Terminology Bank. Although the Study Group's report has not yet been released, it is already known that, if necessary, we could add about 50 terminals to the present network in the next two fiscal years (1979-80 and 1980-19) without greatly increasing the response time and the time required for processing the data. Further, there is every indication that the recommendations of the Study Group will not provide for changes in the basic criteria set out long ago for outside clients who wish to obtain terminals: since the Terminology Bank considers itself a large network of collaborators, those who are interested in obtaining a subscription must agree to contribute to the development of the bank and to the screening of the various files.

At present most translators (89%) in the Translation Bureau have access to a network terminal which, if not directly within their unit, at least is located in the building in which they work. To provide terminal to 100% of translators, it would be necessary to install more than 30 additional terminals, most of which would be in geographically isolated units with only a few translators. Our immediate objective is therefore not necessarily to connect these services to the bank by means of a terminal. Each request for a terminal must be considered on the basis of real needs. Those who do not have acess to a terminal can make use of the Terminology Bank's services by requesting off-line retrieval of data or retrieval by parameters, or by contacting the TIS* units. Off-line retrieval here means that printouts are provided in response to requests for batch retrieval by means of a high-speed printer. We try to limit the waiting time to less than 24 hours for this type of printout. Retrieval by parameters here means that listings are provided in response to requests for data by field, source or author parameters or a combination of all three. The average cost of producing a complete terminological record using the retrieval by parameters system is approximately $0.07 cents, compared with approximately $0.01 for a record showing only the entry terms and the subject field.

Cost to clients

There is no such thing as a paying subscription to the Terminology Bank. The only payment accepted, at least for the time being, is in the form of terminology and comments on the records already in the bank. The terminology in

---

* Terminological Information Service.

the bank is in theory available to anyone who wants it, provided that the individual is willing to pay the "operating costs", that is, provided he installs his own terminal, obtains his own telephone line and pays for the computer time. In other words, the overhead expenses (those related to developing the system, research and data processing) are paid by the Translation Bureau, but the user must assume responsibility for the variable expenses.

For information purposes, the items required and costs* involved in obtaining a connection to the Terminology Bank are:

(a) terminal:
- rental per month: $150
- installation cost: $50
- purchase: $2000
- installation cost: $75

(b) telecommunication line: $7.50 an hour (Datapac network)
$6.50 an hour (Ottawa-Hull network)

(c) computer time: there is no set hourly rate since it depends on the computer time actually used - statistics would seem to indicate that the rate is approximately $8.00.

We are now studying the possibility of offering off-line retrieval and retrieval by parameters, subject to certain conditions, to organizations and businesses which cannot meet our criteria for obtaining terminals or whose needs do not justify connection to the network. The cost of retrieving data would have to be paid by the user in these cases as well.

The terminological data classification system

The Terminology Directorate has been working for almost two years to develop a terminological data classification system. Intensive research has shown that none of the existing classification systems can be used in its present form for indexing and retrieving terminological data from the data base. The reason is simple: terms cannot be classified in the same way as documents. The Classification Group is continually consulting document classification systems, but it has to alter, delete or add descriptors and subject headings to such an extent that the introduction of a new system must be considered.

The current project represents a major undertaking. Though it is still three or four years from completion, a very important step has just been taken. This was a document prepared in English and French by the Group,

---

* These are current costs, which are subject to change.

giving a description of the subject fields. After reviewing approximately 80 subject fields, the Group retained 55; these will include all the terminological data which can be stored in our bank. These subject fields thus become the main divisions of the new classification system. The document is not, however, designed as a tool for classifying terms and retrieving them from the terminology bank. Division by subject area, no matter how well conceived, is still likely to be understood in different ways by different users. It proved necessary to seek a tool offering greater likelihood of accuracy; we finally chose the classification tool known as the thesaurus.

A thesaurus is a structured and controlled vocabulary for classifying documents or terminological data. Its main concern is to establish descriptors (key words used for classification) as accurately as possible in order to avoid the danger of ambiguity in the natural language. To this end, a thesaurus, in its structured alphabetical list, specifies the semantic range of the desciptors by setting out their hierarchical and associative relationships and their synonymy. If we add to this structured alphabetical list a section incorporating a scope note for each descriptor, we may assume that the users of the thesaurus will interpret the descriptors in a very similar if not an identical way. When we are loading thousands or even hundreds of thousands of terms, it is essential to be able to rely on a classification tool providing this assurance of accuracy. Term searching, whether by term of by subject area (vocabulary, glossary) will not fail to produce results, since the scope of the descriptors used to classify terms will be well defined and uniformly understood by the users.

The Group has so far constructed a thesaurus related to geology. Trials of this thesaurus by two terminologists have indicated a very encouraging degree of accuracy in the classification of terms. The Group now has to carry out the second and final stage of its mandate: the construction of a thesaurus embracing the other 54 selected subject fields. This macrothesaurus will be known as the TERMI thesaurus.

Querying

The Termium system was designed expressly to meet the needs of translation and to provide immediate responses to specific queries - queries regarding specific terms and not complete vocabularies.

The user may at present query the data base using one of two methods: on-line and off-line. The result of an on-line query is the immediate receipt of one or more records in response to the request.

When the user makes an off-line query, he does not wait for the computer to transmit the records. The responses are sent to him later in the form of a listing as soon as they emerge from the high-speed terminal.

When the program was designed, one of the chief concerns of the systems analysts was to reduce the response time to a minimum and also make the querying procedure as simple as possible. All querying is by means of a pair the first element of which is an indication of the source language and the second, the term or terms requested in that language. In order to limit term searching, the user may, if he so wishes, specify a subject field. If he does not, the first response is a subject field distribution list for the entry term. This subject field distribution list, by indicating the number of records in each subject field, allows the user to make a selection. All that remains is to specify the subject field or fields required. If the subject field was specified in the request, this step is omitted and the user receives all the records for the term in question immediately.

Records produced by either method include the following main elements:

- the English and French entry terms
- the source code
- definitions of the English and French contexts
- the codes for the subject fields
- the acceptability rating, coded or written out.

Attached is more detailed explanatory diagram of the three types of record currently in the bank. The BTUM (Banque de terminologie de l'Université de Montréal) records were acquired from the University at the same time as the software; they are relatively complete. The TDB (Terminology and Documentation Branch) records, are those collected in the translation sections in 1976. The Termium II records include all the elements most suited to the needs of translators and terminologists. New records added to the bank are displayed in this manner.

A user making an off-line query may, if he wishes, obtain a vocabulary rather than complete records. This is produced as a list of alphabetized or non-alphabetized terms, in English or French, with the equivalences for each term in the other language. The printout also includes an acceptability rating written out and the codes for the subject fields.

Our system makes available off-line printouts to meet individual needs. It is possible to obtain printouts (complete records or vocabularies) arranged according to subject field code, author or source, or any combination of these three elements. It is also possible to retrieve the records by using the record identification number. Other programs will be developed to retrieve the information from the data base in other ways.

Languages

In its present form, the system can process records in any language that use the Latin alphabet. However, the single-concept rule governing the terminology file means that only one record per concept can be loaded into it. At present, each concept appears in two languages on each record, which bears an identification number. It would theoretically be possible to load a term in a third language using a half-record, but only if an a=b=c relationship existed between the concepts designated by the equivalent terms. On the other hand, where there is a complete equivalence between a and b, but a whole or part equivalence between b and c, it would be necessary to prepare two other records and indicate the relationship between a and c on one and between b and c on the other. However, this procedure is not permitted at present. It would be necessary to restrict the single-concept rule to pairs of languages and authorize the loading of as many records as there are pairs of languages. It should also be possible to consider establishing one record per language, each with its own identification number, rather than by pair of languages, but a mean would have to be found to indicate the relationship of the record in language a to the records in languages b, c and so forth for the same concept. Instead, it might be advisable to establish a metalanguage that would express all the concepts designated by all the languages in question.

Processing languages written in an alphabet other than Latin, particularly Cyrillic, does not present a programming problem. The systems analysts already know how to do so; only questions of priority are preventing us from including Russian in the date base.

The problem posed by the single-concept rule does not affect the automatic dictionary or the research file. Therefore it is possible to load records in all languages into them but, here again, there is a question of priority.

Variability of fields

As we mentioned above, the first unique characteristic of Termium II is the creation of an updating system that makes it possible to correct records in the memory or to add additional information on a daily basis. Since some of the records in the memory are provisional only, this is an essential feature: we have decided to store all partial solutions as we discover them or as our clients pass them on to us.

A second important characteristic of our permanent file is the absolute variability of the length and number of fields on the record. In designing our record we were very aware of the need to exchange terminological information on tapes or other computer mediums, particularly with the Quebec Office de la langue française, the Commission of European Communities, the Federal Republic of Germany, the Siemens firm, and others. Since terminological research is very expensive, we must endeavour to avoid duplicating work. If, as we hope, terminology exchanges become systematized, universality and compatibility between terminology banks will be achieved that much more quickly. For our part, this is particularly important in the case of such foreign languages as German, Russian, Danish and Dutch, all of which we need, but for which we will never have a large-scale research service. Instead, we will have to rely on foreign sources to supply data in these languages.

Present and future terminology banks are not all designed to meet the same needs, and their data are not necessarily set up using the same fields. Even if the fields have the same name, the information placed in them often varies from one bank to another.

Since the length and number of fields on our record can be varied, we can set up new fields for items on other records that do not appear on ours. Similarly, we can leave other fields on our record blank, if this is required. In this case, the record would be incomplete according to our criteria, but we would complete it as we obtained more information, using the procedure described above.

The input document

The basic model of the record was designed to resemble as closely as possible the presentation of the information on a screen 80 characters wide by 16 lines high. Although the length of all the fields may vary, we decided that it would be advisable to limit the length on the input document to simplify the preparation, checking and loading of the record. The record we are using can usually accommodate the necessary information; an overflow page is rarely required.

This format, which is more or less the basic model, is to be used by our terminologists, who should of course try to make the information as complete as possible.

Moreover, since we are ultimately relying to a large extent on the contributions of translators in the Bureau, and since the latter have little inclination to complete terminology records in full (because of the time factor and other considerations), we have also developed another document more suited to their needs. Without changing the method they have always used, translators can now use 3" x 5" (75 mm x 125 mm) records to submit information to the bank at the same time as they place it in their personal files. These "mini-records" are filled in by translators, who keep a copy and submit the original to a terminologist.

We have provided that various versions of the record may be used for various purposes. Our goal is to assemble terminology from all areas in which it is used, while disrupting existing procedures as little as possible. To do this, we will attempt to adapt our input document to become part of the terminology production system as smoothly as possible. Other versions will always be compatible with the basic terminology record. And the fields will always contain the same information.

Documentation Interface

The Terminology Bank and the bibliographic file interface through the source code. Under the present system, the source code is used to identify the terminological source and the bibliographic reference. The source code file contains all codes assigned systematically to documentation acquired by the Documentation Directorate for the Translation Bureau's system of libraries and reading rooms, as well as codes assigned to documents from sources outsid the Bureau which are used for terminological purposes. This file ensures tha each code is assigned to only one document, and makes it possible to decode any code assigned, obtain a complete bibliographic entry and locate a documen within the Translation Bureau system and also, in many cases, outside the system.

The Documentation Directorate uses the data processing services of the University of Toronto Library Automation Systems (UTLAS) for cataloguing purposes by contractual agreement. One of the fields of the cataloguing worksheet identifies the source code so that a source code inventory may eventually be produced by computer. This inventory will replace the conventional source code file and will be issued to all translators and terminologists in the Bureau as a microform, probably on microfiche.

In addition, the Documentation Directorate will be able to provide the Terminology Directorate with the Translation Bureau's union catalogue on tape for eventual loading into the bank. This will enable users to obtain not only terminological information but also specialized documentation for terminological research. Such complete interface between terminology and documentation has now become possible.

Conclusion

The purpose of this document is to give a general idea of the terminology bank. It goes without saying that our approach will continue to develop and our methods will become increasingly refined. Consequently, people who wish to keep up to date on new developments should establish contact with the Terminology and Documentation Branch of the Translation Bureau.

Much more detailed documents, which constitute the terminologists' work tools, may be obtained from the Standardization and Liaison Directorate, Terminology and Documentation Branch, Secretary of State Department, Ottawa, K1A 0M5, Canada.

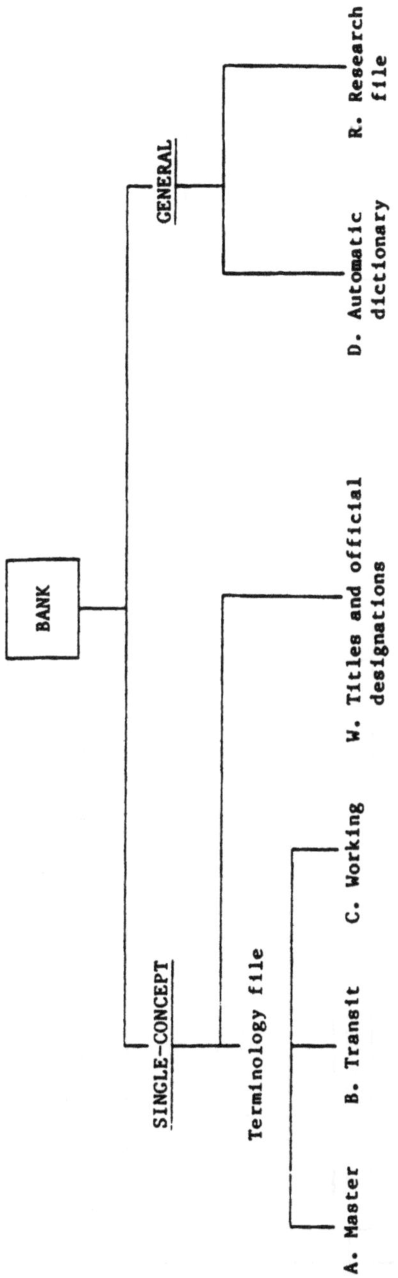
BANK
SINGLE-CONCEPT
GENERAL
Terminology file
A. Master
B. Transit
C. Working
W. Titles and official designations
D. Automatic dictionary
R. Research file

ANNEXE 2

DIRECTION GÉNÉRALE DE LA TERMINOLOGIE ET DE LA DOCUMENTATION

Schéma explicatif de la fiche

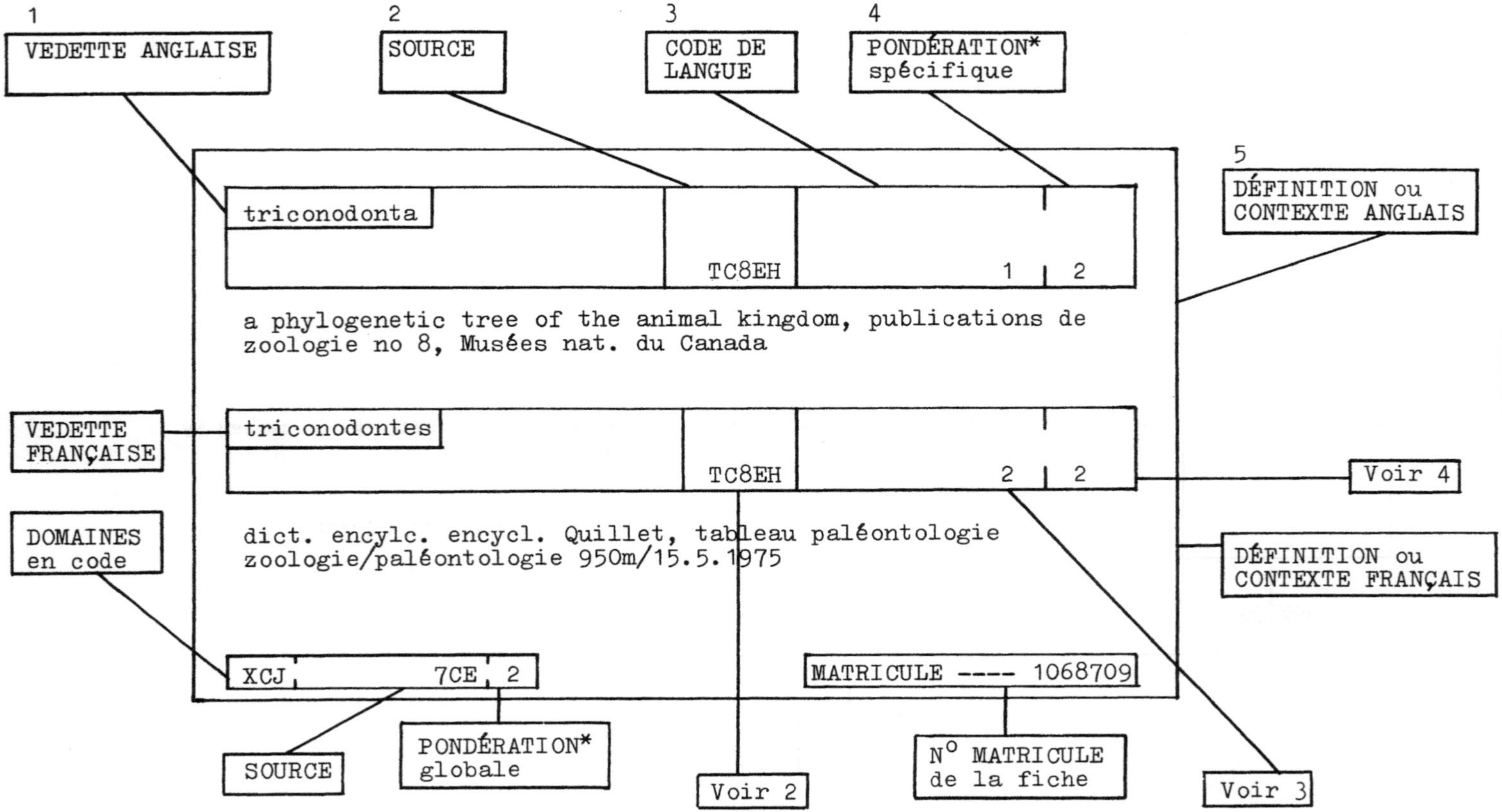

* Lors du classement, toutes les fiches DGTD ont reçu le code de pondération 2 (pondération spécifique

Secrêtariat d'Etat
Direction de la Terminologie
Ottawa, Canada

6.6.12 DESCRIPTION DES DOMAINES

Document de travail à l'usage du groupe de classification

Avant-Propos

Nous vous présentons, à titre d'exemple, la description de six domaines, trois en français et trois en anglais. Le groupe a procédé à la description de cinquante-cinq domaines (voir la liste des domaines); ce document de travail servira de guide pour l'établissement du thesaurus TERMI. Un seul domaine, celui de la Géologie, se présente à nous sous la forme d'un thesaurus mais l'équipe de la classification s'attaquera sous peu à l'établissement du macrothesaurus bilingue (français-anglais) qui couvrira les cinquante-quatre autres domaines. Ce macrothesaurus, nous l'appelons le thesaurus TERMI afin de signaler qu'il entend être un outil de classement des données terminologiques. En lisant la description des six domaines que nous vous présentons, vous aurez une idée du contenu de ces domaines mais il faudra vous rappeler que le language du thesaurus pour chacun de ces domaines différera de celui de leur description.

LISTE DES DOMAINES

1. ADMINISTRATION PUBLIQUE
2. AGRICULTURE
3. ART ET SCIENCE MILITAIRES
4. ARTS
5. ASTRONAUTIQUE
6. ASTRONOMIE
7. BIOLOGIE
8. CHIMIE
9. COMMERCE
10. CONSTRUCTION
11. DROIT
12. ÉDUCATION
13. ÉLECTROTECHNIQUE
14. ÉNERGIE
15. ENTREPRISES COMMERCIALES
16. FORESTERIE
17. GÉNIE MECANIQUE
18. GÉOGRAPHIE
19. GÉOLOGIE
20. HISTOIRE
21. INDUSTRIES MANUFACTURIÈRES
22. INFORMATIQUE
23. LOISIRS
24. MANAGEMENT
25. MATERIAUX DE CONSTRUCTION
26. MATHÉMATIQUES
27. MÉDECINE
28. MÉDECINE VÉTÉRINAIRE
29. MÉTALLURGIE
30. MÉTROLOGIE
31. MINES
32. OCÉANOGRAPHIE
33. PÂTES ET PAPIERS
34. PÊCHES
35. PHARMACOLOGIE
36. PHILOSOPHIE
37. PHOTOGRAPHIE
38. PHYSIQUE
39. POLLUTION
40. POSTES
41. PSYCHOLOGIE
42. RELIGION
43. SCIENCE CRIMINELLE
44. SCIENCE DE L'INFORMATION
45. SCIENCE ÉCONOMIQUE
46. SCIENCE POLITIQUE
47. SCIENCES DE L'ATMOSPHÈRE
48. SCIENCES DU LANGAGE
49. SÉCURITÉ
50. SERVICES SOCIAUX
51. SOCIOLOGIE
52. SPORTS
53. TÉLÉCOMMUNICATIONS
54. TRANSPORTS
55. TRAVAIL

BIOLOGIE

Terminologie touchant les phénomènes communs à tous les êtres vivants ainsi que les phénomènes biologiques propres à chaque catégorie d'êtres vivants (genre humain, animaux, végétaux). Ce domaine comprend:

- l'anatomie
    - l'anatomie humaine
    - l'anatomie animale
    - l'anatomie végétale
    - l'anatomie comparée
    - l'embryologie
    - etc.
- la biochimie
    - l'histochimie
    - l'endocrinologie
    - la microchimie
    - les vitamines
    - etc.
- la biogenèse (évolution de la vie)
- la biogéographie
    - la phytogéographie
    - la zoogéographie
- la biologie moléculaire
- la biométrie
    - les biomathématiques
    - la biostatistique
- la biophysique
    - la biomécanique
    - la bionique
    - l'électrophysiologie
    - la radiobiologie
    - etc.
- la botanique
    - la pathologie végétale
    - la palynologie
    - la phytosociologie
    - la taxonomie
    - etc.
- la cryobiologie

ASTRONOMIE

Terminologie touchant les sciences qui étudient tous les corps extérieurs à la Terre et plus généralement tout l'espace dans lequel ils se trouvent. Ce domaine comprend:

- l'astronomie théorique

- la cosmologie
    - la cosmogonie
    - la cosmographie

- l'observation astronomique
    - les observatoires
    - les techniques
    - les instruments astronomiques et astrophysiques

- les astres
    - les lunes
    - les planètes
    - les comètes
    - les galaxies
    - les systèmes extragalactiques
    - etc.

- les météores et les météorites

- l'astronomie de position

- l'astrophysique

- l'astrogéologie

- l'astrographie

- la mécanique céleste

- la radioastronomie

- l'astronomie corpusculaire

- l'astrologie

Sont exclues: l'astrométrie, l'actinométrie (MÉTROLOGIE); l'astrionique (ASTRONAUTIQUE).

SCIENCES DU LANGAGE

Terminologie touchant l'étude et l'analyse des langues et du langage humain. Ce domaine comprend:

- les langues
    - les sortes de langue et langage (ex.: langue morte, dialecte)
    - la classification des langues

- la linguistique
    - les théories linguistiques
    - la linguistique descriptive
        - la phonétique et la phonologie
        - la grammaire
        - la morphologie
        - la syntaxe
        - la sémantique
    - la linguistique diachronique
        - la reconstruction interne
    - la linguistique contrastive
        - la traduction et l'interprétation

- les phénomènes paralinguistiques
    - les modes de comportement (ex.: tic, geste)
    - les vocalisations (ex.: timbre, qualité de la voix, ronflement)
    - le langage gestuel

- la lexicographie et la lexicologie

- la terminologie

- les systèmes d'écriture
    - les alphabets
    - la ponctuation

- la psycholinguistique
    - l'acquisition de la langue maternelle
    - les troubles de la communication verbale

- la socio-ethno-linguistique
    - la géographie linguistique (dialectologie)

- la littérature (orale et écrite)
    - la stylistique
    - la rhétorique
    - les genres littéraires
    - les modes littéraires
    - la critique littéraire

- la linguistique mathématique (ou computationnelle)
    - la traduction automatique
    - la statistique lexicale

- l'orthophonie

Est exclue: la pédagolinguistique (ÉDUCATION).

- l'écologie
    - l'autoécologie
    - la démécologie
    - la synécologie

- la génétique

- l'immunologie

- la microbiologie
    - la bactériologie
    - la parasitologie
    - la virologie
    - etc.

- la nutrition

- la physiologie
    - le métabolisme
    - la physiologie du stress
    - le système nerveux
    - etc.

- la recherche biologique (y compris la pêche pour fins de recherche biologique)

- la zoologie
    - l'acarologie
    - l'entomologie
    - la taxonomie
    - etc.

Secretary of State
Terminology Directorate
Ottawa, Canada

6.6.13 SUBJECT-FIELD DESCRIPTIONS

Working Document of the Classification Group

LIST OF SUBJECT-FIELDS

1. AGRICULTURE
2. ARTS
3. ASTRONAUTICS
4. ASTRONOMY
5. ATMOSPHERIC SCIENCES
6. BIOLOGY
7. CHEMISTRY
8. COMMERCE
9. COMMERCIAL ENTERPRISES
10. CONSTRUCTION
11. CONSTRUCTION MATERIAL
12. CRIMINAL SCIENCE
13. ECONOMICS
14. EDUCATION
15. ELECTRICAL ENGINEERING
16. ELECTRONIC DATA PROCESSING
17. ENERGY
18. FISHERIES
19. FORESTRY
20. GEOGRAPHY
21. GEOLOGY
22. HISTORY
23. INFORMATION SCIENCE
24. LABOUR
25. LANGUAGE
26. LAW
27. LEISURE ACTIVITIES
28. MANAGEMENT
29. MANUFACTURING INDUSTRIES
30. MATHEMATICS
31. MECHANICAL ENGINEERING
32. MEDICINE
33. METALLURGY
34. METROLOGY
35. MILITARY ART AND SCIENCE
36. MINING
37. OCEANOGRAPHY
38. PHARMACOLOGY
39. PHILOSOPHY
40. PHOTOGRAPHY
41. PHYSICS
42. POLITICAL SCIENCE
43. POLLUTION
44. POSTAL SYSTEMS
45. PSYCHOLOGY
46. PUBLIC ADMINISTRATION
47. PULP AND PAPER
48. RELIGION
49. SAFETY
50. SOCIAL SERVICES
51. SOCIOLOGY
52. SPORTS
53. TELECOMMUNICATIONS
54. TRANSPORTATION
55. VETERINARY MEDICINE

AGRICULTURE

Terminology pertaining to the production and primary processing of plants and animals useful to human beings. This includes:

- animal production
    - different species (e.g. swine, poultry, cattle)
    - mating systems (e.g. inbreeding, crossbreeding)
    - breeding methods (e.g. artificial insemination)
    - etc.

- plant production
    - types of breeding
    - breeding methods
    - etc.

- field crops
    - forage crops
    - grain crops
    - pastures
    - etc.

- horticultural crops
    - fruit
    - vegetables
    - etc.

- crop farming methods
    - tilling
    - cultivation
    - harvesting
    - dryland farming
    - etc.

- specialty crops
    - honey
    - maple syrup
    - etc.

- dairy farming
    - methods
    - etc.

- soil management
    - fertilizing
    - irrigation
    - etc.

- farm equipment
    - machines
    - implements
    - etc.

- farm structures

- agricultural engineering

- farm management
  - farm personnel
  - organization
  - types of farms
  - etc.

- agricultural economics
  - marketing
  - farm policy
  - etc.

Excluded are: botany (BIOLOGY); zoology (BIOLOGY); sílviculture (FORESTRY); pedology (GEOLOGY); aquaculture (FISHERIES); slaughterhouses (MANUFACTURING INDUSTRIES).

ASTRONAUTICS

Terminology pertaining to space travel and exploration. This includes:

- spacecraft (manned and unmanned, including satellites and probes, etc.)
    - types of spacecraft
    - parts of spacecraft (e.g. lunar excursion module, astrionics, nose cone)
    - craft assembly
    - performance trials
    - specially developed fuels and power cells
    - functioning and malfunctioning
    - etc.

- ground-based facilities and operations
    - launch complex
    - tracking network
    - mission control
    - recovery

- space equipment (other than that which is part of the vehicle)
    - space suits
    - tools for sample collection
    - etc.

- space environment
    - environmental conditions (e.g. weightlessness, solar wind)

- space crew
    - training exercises and facilities
    - etc.

- spaceflight
    - types
    - stages
    - characteristics
        - length
        - altitude
        - speed
        - trajectory
        - etc.

- UFO's

Excluded are: psychological and physiological effects of spaceflight conditions (MEDICINE).

METROLOGY

Terminology pertaining to weights and measures, that is, measurement indicated and controlled by measuring instruments. This includes:

- measurands (e.g. motion, pressure, flow)

- units
    - nomenclature (e.g. metre, watt, radian)
    - classification
    - multiples and submultiples of units (e.g. mega, nano)

- instruments
    - measuring tools
    - measuring machines
    - maintenance
    - performance
    - etc.

- measurement techniques (e.g. direct measurement, measurement by deviation, telemetry)

- mechanics of measurement
    - measuring processes (e.g. calibration, transduction)
    - measuring facilities
        - calibration centers
        - metrological stations
        - laboratories
        - etc.

- historical metrology (obsolete measuring units, systems, instruments, and methods)

Excluded are: bibliometrics (INFORMATION SCIENCE); ergonometrics (MANAGEMENT); econometrics (ECONOMICS); psychometrics (PSYCHOLOGY).

Secretary of State
Terminology Directorate
Ottawa, Canada

6.6.14 LES MÊTHODES DE TRAVAIL DU GROUPE DE CLASSIFICATION/ THE WORK METHODS USED BY THE CLASSIFICATION GROUP

OBJECTIF DU GROUPE

L'établissement du thésaurus TERMI à l'aide du document rédigé par le groupe et intitulé: "La description des domaines".

OBJECTIF DE TRAVAIL

1. Le travail doit être mené dans les deux langues officielles.

2. Se documenter sur la façon d'établir un thésaurus. Faire les lectures nécessaires afin de se familiariser avec le domaine choisi. Examiner les dossiers déjà constitués sur le domaine en question.

3. Faire une recherche en banque sur le domaine qu'on traite. Examiner comment les termes sont classés.

4. Augmenter la bibliographie du domaine en s'en tenant principalement aux thésaurus, aux dictionnaires et aux glossaires du domaine et des sous-domaines. Etablir un fichier portant sur la documentation disponible et les sources utilisées..

5. Déterminer les spécialistes qui pourraient faire partie d'un comité de révision.

6. Établir une liste de descripteurs candidats en provenance de thésaurus déjà existants et de systèmes de classification.

7. A partir de la liste des descripteurs candidats, choisir ceux qui pourraient servir de descripteurs du second niveau, constituant ainsi les sous-divisions immédiates du grand domaine à l'étude. Cette seconde liste peut changer au cours de la période de recherche.

8. Fouiller en profondeur chacun des descripteurs du second niveau afin de déterminer s'ils renferment assez de terminologie pour être retenus. Il faut aussi déterminer si les descripteurs du second niveau ont besoin d'être sous-divisés ou non; en d'autres mots, déterminer si l'on doit créer des descripteurs du troisième niveau ou non. Pour ce faire, il faut:

    1) consulter et analyser les glossaires et les thésaurus

    2) passer en revue les dictionnaires, les manuels, les encyclopédies et autres documents afin de reconnaître les termes spécifiques au domaine.

    3) consulter des spécialistes

   (voir document ci-joint portant sur la géophysique)

9. Identifier les descripteurs synonymes et quasi-synonymes.

10. Établir un fichier de définitions pour chacun des descripteurs du domaine.

11. Choisir le descripteur préférentiel et établir la hiérarchie (relations génériques, spécifiques, et synonymes) pour chacun des descripteurs. Consigner les renseignements sur fiches selon le format utilisé dans "Production of a Terminology Thesaurus".

12. Soumettre la recherche sur chacun des descripteurs au comité de spécialistes. La révision devrait porter sur:

    1. la liste des descripteurs candidats examinés

    2. les descripteurs retenus

    3. les hiérarchies établies

    4. les descripteurs synonymes

    5. les définitions des descripteurs

    6. les questions soulevées par la recherche

13. Mettre au point la structure et établir des liens avec le reste du système en identifiant les domaines associés.

14. Mettre à l'essai le thésaurus en utilisant un échantillon de fiches du domaine en question et apporter les modifications qui s'imposent.

ADMINISTRATION

Remplir au besoin les tâches administratives suivantes: rédige les comptes rendus, les rapports mensuels; assiste aux réunions de section et de division; apporte sa contribution aux rapports; aide à la préparation de la présentation du système de classification; participe activement à l'élaboration de nouvelles méthodes de travail.

PERFECTIONNEMENT

1. Signaler au groupe tout matériel se rapportant à la classification et lire le matériel que l'on fait circuler à l'intérieur du groupe.

2. Assister aux séminaires, lectures, congrès, etc.

GROUP OBJECTIVE

Construction of the TERMI thesaurus based upon a working document written by the group and entitled "The Subject-Field Descriptions".

WORK OBJECTIVES

1. All work must be carried out in both French and English.

2. Do background reading to become acquainted with thesaurus construction. Do background reading to become acquainted with the field. Examine the file previously established on the main subject-field in question.

3. Make a request for the records in the bank on the subject-field. Examine how the terms are classified.

4. Expand the bibliography of the field, focussing mainly on thesauri, dictionaries and glossaries of the field and subfields. Establish a card file of the documentation available and sources used.

5. Identify specialists in the field who might participate in a committee of specialists for the classification group.

6. Establish a list of candidate descriptors from existing thesauri and classification schemes.

7. From the list of candidate descriptors, select those which could act as second level descriptor which would form each immediate branch of the main subject-field. This second list of candidate descriptors can, and probably will, vary throughout the research period.

8. Research each branch to determine, as thoroughly as possible, whether the second level candidate descriptor generates sufficient terminology to warrant a descriptor and proceed to delimit the scope and the divisions of each branch. This requires:

   1) consulting and analysing glossaries and thesauri

   2) scanning dictionaries, textbooks, encyclopedias and other basic documents for terms

   3) consulting specialists

   to identify some of the terminology and to establish the level of specificity required (2nd, 3rd level etc.) (See attached sheets on Geophysics)

9. Identify synonymous and quasi-synonymous descriptors.

10. Establish a card file of definitions for each subject-field descriptor.

11. Choose the preferred descriptor and establish the hierarchy (broader, narrower and synonymous relationships) for each branch. Record the information on cards according to the format outlined in the "Production of a Terminology Thesaurus".

12. Consult field specialists or submit the research of each branch to the committee of specialists established for this purpose. The research should include:

    1) the list of candidate descriptors examined

    2) the descriptors retained

    3) the hierarchies established

    4) the synonymous descriptors

    5) the definitions of the descriptors

    6) the questions arising from the research

13. Finalize the structure and establish the links with the rest of the system by identifying related fields.

14. Test the thesaurus with a sample of records in the field and make any necessary adjustments.

ADMINISTRATION

Perform the following administrative duties: write minutes, prepare monthly reports, attend section and division meetings, contribute to reports, help prepare presentations of the clasification system and participate in elaborating and perfecting work methods.

DEVELOPMENT

1. Bring to the attention of the group any material that is relevant to classification and read that which is circulated by other group members.

2. Attend seminars, lectures and the like.

7 OTHER STATEMENTS AND OBSERVATIONS

The audience was asked whether any data bank would wish to make additional statements or observations. To this suggestion by the chairmen

- GOSSTANDART stated that it offered 60.000 term records of standardized terms for exchange and that it welcomed the setting up of TermNet. For this purpose it advocated the harmonization of classification schemes and data records.
- Mr. Landry stated that he hoped Unesco would support the network and would impress upon member states the need of this work.
- Mr.Löhner replied that it would, therefore, be more appropriate for conference members to request their respective governments to direct Unesco's attention to the importance of terminology.
- Mr. Goetschalckx observed that the Nairobi Convention on Translators contained a recommendation for national governments to develop terminology centres.

8 Recommendations

The participants of the First International Conference on Terminological Data Banks recommended that Infoterm should be requested

1 to organize such conferences on a regular basis, preferably with emphasis on particular themes;

2 to set up a task force (working group) with the two-fold objective of:

a) gathering information about the organization, structure and content of data stored by existing terminology banks, and regularly distributing this information to all interesting parties with a view to facilitating the exchange of data; such inventories would have to contain a minimum set of items of information such as subject field, language(s), availability of data to third parties and the conditions under which these data are offered;

b) supplying terminology banks in the process of implementation or development with information and guidance as and when requested; (Rondeau)

3 to undertake or to arrange for a comparative study to be undertaken of the categories (data elements) employed in different term records for ordering and identifying terminological data; (Gorm-Hansen)

4 to undertake or coordinate the development of guidelines for the establishment of terminological data banks; (Arnold)

5 to undertake or to direct a survey of the solutions which have been found by existing terminological data banks to overcome user reluctance towards working with computer systems and especially with screens. (This request is based on the awareness that the relationship between

man and machine is one of the many problems which have arisen from the use of terminological data banks.); (Bachrach)

6 to coordinate efforts with a view to making proposals towards the harmonization of the methods of coding source material; (Al-Kasimi)

7 to collect and distribute information about the subject classification systems used by existing terminological data banks, as a first step towards achieving compatibility among the subject classifications in use. (Dovbenko and Zdorovov).

9 Closure of the Conference

Mr. Felber thanked the chairmen of the sessions and the Rapporteur, Mr. Sager, for their assistance and expressed the hope that Unesco would continue to sponsor the work of Infoterm.

10 APPENDIX

10.1 Krommer-Benz, M. Co-operation in TermNet.
Analysis of Questionnaires

10.2 Felber, H. A Plan for Data Management within
TermNet

10.3 TermNet Follow-up 2

M. Krommer-Benz
Infoterm
Vienna, Austria

10.1 ANALYSIS OF QUESTIONNAIRES CONCERNING CO-OPERATION IN TERMNET

1. BACKGROUND

On the basis of a recommendation expressed by participants of the First Infoterm Symposium on International Co-operation in Terminology (Vienna, April 9 - 11, 1975) a study had been prepared by Infoterm on the establishment of a network of terminology information and documentation (TermNet). After evaluation of this study by a panel of experts convened by Unesco (Vienna, October 19 - 21, 1977) and subsequent revision and amendments the panel recommended to proceed immediately with the implementation of TermNet.

In compliance with this request a communication was distributed in spring 1978 to a limited number of likely members inviting them to join TermNet. This communication consisted of the amended study on the development of a network for terminology documentation and information, a questionnaire requesting information on fields in which collaboration was desired, etc. Deadline for the return of the questionnaire had been set for May 31st, 1978; it had to be extended, however, to allow for detailed discussion within a number of large international organizations. Analysis could thus be completed by September 1979.

2. RESULTS

2.1 Summary of the evaluation

Number of questionnaires distributed:	180	
Number of questionnaires returned:	57	
of these positive:	46	(one with certain reservations as to the seeming priority given to the English language)
negative:	11	

Negative answers:

In general respondents refrained from completion of the questionnaire because they felt that terminology work was not or no longer of primary interest to them. They, therefore, did not feel a need for joining TermNet even though they wanted to be kept informed about its development.

5 of the negative responses were sent in by international organizations composed of member bodies carrying out particular tasks. Thus, these international organizations themselves did not pursue terminological work per se; if anything in this field was undertaken by them it was done by ad-hoc commissions or committees only to an extent necessary for fulfilling the tasks set out for the international organization by its statutes.

The remaining 6 negative answers were from national centers specializing in a particular field (such as engineering) or from research founding bodies. They were not involved in terminology work at all except in a very marginal sense.

Both groups of respondents could not envisage participation in an endeavour such as TermNet since they could not contribute to TermNet in any way.

2.2 Present activities

With respect to present activities 38 respondents focused on the elaboration of terminologies, 38 on documentation of terminologies, with a majority, namely 21 applying manual processing of information, 12 using both, manual and computerized methods for data recording and processing, 10 organizatuons carrying out processing via computers only.

At present, 33 organizations participate already in a co-operative terminology scheme, basically either language (e.g. CILF-OLF) or region oriented (e.g. NORDTERM).

2.3 Collaboration within TermNet with respect to certain projects

A great number of institutions are interested in collaboration with respect to

BT 2	(Int. bibliography of standardized vocabularies)	35
BT 3/4	(Int. Bibliography of mono- and multilingual specialized vocabularies)	40
BT 1/6	(Int. bibliography of theory and methods of terminology)	33
BT 7	(Int. bibliography of periodicals related to terminological matter)	32
BT 5	(Revision of the "World Guide to terminological activities")	41

Subject areas in which collaboration is being sought naturally coincide with those already embarked upon.

The table below underlines the fact that activities of the majority of respondents (nearly 50 %) practically span all subject areas (UDC 0/9).

Co-operation desired in subject field(s)

0/9	All fields / Tous les domaines	18
0/6	Philosophy, Social Sciences. Law. Mathematics. Natural Sciences. Applied Sciences. Medicine. Technology / Philosophie. Sciences sociales. Droit. Mathématiques. Sciences naturelles. Sciences appliquées. Médecine. Technique	2
3/6	Social Sciences. Mathematics and Natural Sciences. Applied Sciences. Medicine. Technology / Sciences sociales. Mathématiques Sciences naturelles. Sciences appliquées. Médecine. Technique	3
5/7	Mathematics and Natural Sciences. Applied Sciences. Medicine. Technology. The Arts. Architecture. Entertainment / Mathématiques. Sciences naturelles. Sciences appliquées. Médecine. Technique. Art. Architecture. Divertissements.	4
001.4.001	Theory of Terminology	1
002	Documentation / Documentation	3
007	Information. Communication and Control Theory. General Cybernetics / Information. Communications. Cybernetique en général	4
008	Civilization. Culture / Civilisation. Culture	1
02	Librarianship / Bibliothéconomie	1
3	Social Sciences. Law Administration / Sciences sociales. Droit. Administration	1

	Co-operation in subject field(s)	
33	Economics / Economie	1
34	Law. Legislation / Droit. Législation	1
35	Public Administration. Military science/ Administration publique. Sciences militaires	1
37	Education / Education. Enseignement	2
528.9	Cartography / Cartographie	1
54	Chemistry / Chimie	1
551.5	Meteorology / Météorologie	1
620.1	Materials testing / Essai des matériaux	1
621.3	Electrical Engineering. Telecommunication / Electrotechnique. Télécommunications	2
621.39	Telecommunication. Telecontrol / Télécommunication et télécommande	1
621.56/.59	Refrigeration technology / Technique du froid	1
63	Agriculture. Forestry. Plants / Agriculture. Sylviculture. Plantes	2
681.2	Measuring instruments	1
681.3	Data Processing. Machines and Equipment. Automatic Data Processing, etc. / Traitement de l'information. Appareils et matériel. Traitement automatique	2
79	Entertainment. Sport /Divertissements. Jeux	1
91	Geography / Géographie	1

2.4 Terminology banks

- 14 respondents dispose of a terminology bank (a computerized terminology file),
- 14 intend to develop such a bank (additional 2 are in doubt)
- 33 are interested in co-operating with other terminology banks. (plus 2 doubtful)

As to proposals (provided by 18 respondents) concerning interconnexion of terminology banks the majority is concerned with questions of information categories, exchange formats, classification systems, and all other matters of standardization for exchange purposes.

In addition, it was felt that legal and financial implication should be studied in detail even though a number of these problems could only be dealt with after the initial phase of the TermNet operation.

Close collaboration with other networks, such as ISO-Net, Euronet, etc. should also be assured.

2.5 Training in terminology

- 33 respondents felt that training was needed most in terminology principles,
- 30 in methods of terminological lexicography, and
- 25 in the fundamentations of terminology.

2.6 Contributions to TermNet

All respondents practically expressed a willingness to contribute in kind; two considered it likely to contribute in money, 6 others were as yet undecided.

3. SUGGESTIONS CONCERNING THE STRUCTURE OF TERMNET, ADDITIONAL TASKS TO BE UNDERTAKEN WITH PARTICULAR REFERENCE TO THE ORGANIZATION OF THE EXCHANGE OF INFORMATION.

Abbreviated responses to item 6 of the questionnaire are given below.

- Co-operation of Arab states necessary
- Special efforts to be undertaken concerning application of standardized terminological principles
- Auspices of Infoterm; detailed structure, etc. to be discussed at general assembly in 1979
- Should first be studied by advisory panel
- Specific tasks should be assigned and carried out in accordance with Infoterm directives
- National clearing houses
- Creation of sub-networks; training of terminologists desirable
- Study committee should study the structure of TermNet, to be discussed at a general assembly in 1979
- TermNet newsletter (like "langues et terminologie" de CILF) should be published
- Wider dissemination of Infoterm publications; dissemination of neologism
- Joint preparation of basic lists of terms for specific fields; newsletter to be made available regularly
- Special documentation centres should be able to provide exact information on specialized vocabularies, and even on specific terms
- Delimitation of fields will be difficult
- TermNet members should pay periodical charge. Specific demands should be placed upon TermNet.
- Necessary to stress co-ordination in social sciences otherwise delegate this co-ordination to other body

Dissemination of up-to-date information (such as WG) necessary

Updating of works of reference, such as World Guide would be necessary

ISO/DIS 6156 "Format for terminological/lexicographical Exchange" should be recommended

Exchange of terminological data for purposes of dictionary publications

ANNEXE: Q U E S T I O N N A I R E

concerning co-operation in TermNet

PLEASE WRITE LEGIBLY!
CHECK BOXES, IF PROVIDED!

1. Name of organization (in original language):.........
....

1.1 Acronym: ....

1.2 Name of director: ....

1.3 Name of expert: ....

---

2. Complete address:..
....

---

3 Comments on the Study enclosed: ....
....

3.1 Based on your knowledge and experience does the study reflect the present state-of-the-art of terminology?

YES ☐ NO ☐

If no, please specify ....
....

3.2 Do you have any additional comments on the study? ....
....

---

4. Present terminological activities focus on:

4.1 Preparation (elaboration) of terminologies

YES ☐ NO ☐

Field(s) ....
.................................(please specify)

4.2 Documentation of terminologies

YES ☐ NO ☐

Field(s) ....
....(please specify)

Data recording and processing:

conventional (manual) ☐ ☐ machine-readable

4.3 Do you participate in any terminology co-operative scheme?

YES ☐ NO ☐

If yes please specify which one?.....................
....

---

5. Collaboration within TermNet (For details please see item 6 of "Study on the Development of a Network for Terminology Information and Documentation prepared by Infoterm)

5.1 Subject field(s) in which collaboration within TermNet is sought most (Please specify, e.g. chemistry, oceanography, welding, etc.)........................
....

5.2 With respect to international projects would you be willing to co-operate in:

5.2.1 International bibliography of standardized vocabularies

YES ☐ NO ☐

5.2.2 International bibliography of mono- and multilingual specialized vocabularies (to be prepared by subject field)

YES ☐ NO ☐

5.2.3 International bibliography of theory, methods and organization of terminology

YES ☐ NO ☐

5.2.4 International bibliography of periodicals related to terminological matter

YES ☐ NO ☐

5.2.5 Revision of the "World Guide to terminological activities"

YES ☐ NO ☐

5.3 Terminology banks

5.3.1 Do you have a computerized terminology file (terminology bank)?

YES ☐ NO ☐

If no, do you intend to develop such a bank?

YES ☐ NO ☐

5.3.1.1 Are you (would you be) interested in co-operating with other terminology banks?

YES ☐ NO ☐

5.3.1.2 What are your proposals concerning interconnexion of terminology banks: ....
....

---

6 Based on item 6 of the "Study on the Development of a Network for Terminology Information and Documentation prepared by Infoterm" and the purposes and structure proposed therein what are your suggestions concerning the structure of TermNet or additional tasks to be undertaken with particular reference to the organization of the exchange of information ....
....

7 Training in terminology

In co-operation with Infoterm the International Association for Applied Linguistics (AILA) plans to organize courses for the training of terminologists, terminological documentalists and teachers of terminology (See Infoterm Newsletter 7 enclosed)

According to your opinion training at present is needed most in

Fundamentals of terminology	YES ☐	NO ☐
Terminological principles	YES ☐	NO ☐
Methods of terminological lexicography	YES ☐	NO ☐

---

8 Would you basically be interested in joining TermNet

YES ☐ NO ☐

If no, we should be interested in knowing why not.......

....

....

---

9 Would you be willing to contribute to TermNet

in kind YES ☐ NO ☐

in money YES ☐ NO ☐

Thank you very much for your co-operation.

H. Felber
Infoterm
Vienna, Austria

10.2 A PLAN FOR DATA MANAGEMENT WITHIN TERMNET

DRAFT

1 INTRODUCTION

In autumn 1971 Infoterm was established in Vienna with the assistance of Unesco within the framework of UNISIST (now General Information Programme of Unesco). It was affiliated to the Austrian Standards Institute [1].
At the first Infoterm Symposium on "International co-operation in terminology" in April 1975 networking of terminological activities was discussed [2]. Infoterm was charged with the elaboration of a study on the development of a network for terminology information and documentation. This study was prepared by Infoterm in 1977 [3]. It was discussed, modified and approved by a group of experts convened by Unesco in Vienna in October 1977 [4].
In 1978 an inquiry about the participation in TermNet was made by Infoterm amoung organizations engaged in terminology in all over the world. The result of the analysis of the questionnaires was distributed to those answering the questionnaire [5].

2 TERMNET PROGRAMMES

TermNet comprises three programmes:

Programme 1: Development of the foundation of terminology work (General theory of terminology)

Programme 2: Development of closer co-operation in terminology work (Organization of the creation and elaboration of terminologies)

Programme 3: Development of a network for terminology documentation and information.

The present plan concerns Programme 3 only, the organization of co-operation in collecting, evaluating, recording, processing, storing, exchanging and disseminating of terminological information.

Terminological information comprises the following types of data:

(1) terminological data (such as terms, synonyms, definitions, context, foreign language equivalents, sources, code of quality, etc.)
(2) bibliographical data on terminological literature
(3) data on the management of terminology work, terminology experts, terminology projects, etc.

For terminological data a separate plan will have to be elaborated in co-operation with the terminology banks concerned. This plan comprises only the bibliographical data and the data on the management of terminology work.

3 DATA MANAGEMENT

TermNet (Programme 3) is intended to be a network for decentralized data recording, if possible, in machine-readable form (or at least with unified worksheets), and for data processing by terminology agencies having a computer and being willing to do it on an exchange basis. The network has the purpose that data will be recorded only once, thus duplication of effort is avoided. For this purpose it will be necessary for those participating in TermNet to agree on: who records what kind of data and who processes them.

3.1. Exchange of magnetic tapes

The magnetic tapes on which the data are stored are to be exchanged among those processing units which participate in TermNet. Thus, each processing unit will have a full set of data. Those participating in the recording of data should have the possibility to receive these data on microfiche (or in any other form) in exchange for their contribution of data on worksheets or in machine-readable form.

For the exchange of magnetic tapes the international standard ISO 2709-1973 "Documentation-Format for the bibliographic information interchange on magnetic tape" will have to be observed. Since TermNet functions within the framework of the General Information Programme of Unesco the "UNISIST Reference Manual for machine-readable bibliographic descriptions" (Doc. Unesco SC 74/WS/20) which was published by Unesco will also have to be taken into account wherever applicable.

3.2 Terminological data

The management of terminological data requires a separate plan to be established in co-operation with terminology banks.

3.3 Bibliographical data

The following international bibliographies are to be elaborated and to be kept up to date:

BT 1 International bibliography of terminological papers and books. Present situation:
A pilot project is being undertaken by Infoterm, AILA (University Laval, Québec, Canada) and the Office de la Langue Française (Québec) and other organizations with the aim to store the data on magnetic tape and to publish the bibliography as soon as possible.

BT 2 International bibliography of standardized vocabularies [6].
Present situation:
These data are stored on magnetic tape by Infoterm. The bibliography was published in 1979 as Infoterm Series 2. It is being kept up to date.

BT 3 International bibliography of specialized vocabularies. This bibliography will be devided in various subject groups. Present situation:

Two pilot projects have been started by Inforterm, one with the Economic Commission for Europe (ECE) of UN in the field of building and construction, another with Interconcept (Unesco) in the field of social sciences.

3.4 Data on the management of terminology work

Data with respect to terminology organizations, experts, projects, etc. should be easily accessible for all concerned with terminology. These data are to be recorded decentrally, processed however at a central place. Infoterm intends to care for the central processing of these data, if the necessary funds are being provided.
Infoterm has so far elaborated:

BT 5 World Guide to terminological activities.
Present situation:
The book was published in 1977 as Infoterm Series 4 [7].
A pilot project is being undertaken by Infoterm with the Institute for automatic documentation of the computer centre Graz to find an optimal solution for central processing of these data.

4 DATA ELEMENTS

The information to be recorded, processed or exchanged in TermNet is broken down into data elements.
The definition for data element is as follows:

data element: the smallest piece of information which is used to describe a concept, a document or an organization, an expert or a project, etc.

In order to make an exchange possible the data elements have to be defined and their form unified.
Example for a data definition:

term: Any conventional symbol for a concept.
A term may be a word or a phrase (ISO 1087)

Example for the prescription of data form:

Capitalization of a term at the beginning of an entry indicated that this term begins with a capital letter even when used inside of a sentence (ISO 1951, 1.2.1.2)

Homogenity of data will allow a mixing and publishing of data without post-editing. For the definition of data elements and prescription of form international standards should be used.

4.1. Description of concepts (terminological data)

Working Group 4 "Computational aids for terminology and lexicography" of ISO/TC 37 "Terminology (principles and co-ordination)" prepared DP6156 "Magnetic tape exchange format for terminological/lexicographical records (MATER)". In this draft proposal a comprehensive list of terminological data is given.

4.2 Description of documents

For the unified recording of the bibliographic data of BT 1, BT 2 and BT 3 (see 3.3) Infoterm elaborated lists of data elements and "Draft TermNet Guidelines for recording of bibliographic data" (doc. TermNet 5 (5)-79), which are in line with ISO 690 "Bibliographic references". These Guidelines are at present revised for compatibility and convertibili by UNIBID and will be made available in form of a Manual in the near future.

4.3 Description of organizations and experts engaged in terminology, etc.

For unified recording of data on organizations and experts engaged in terminology, etc. (BT 5) Infoterm elaborated worksheets and "Draft TermNet Guidelines for recording of data of terminology work" (organizations, commissions, experts, projects, etc.)

4.4 Description of the content of documents

A classification system is necessary for the description of the content of documents. Infoterm uses the Universal Decimal Classification (UDC) for BT 2 and BT 3 and a subject code for DT 1, since UDC for terminology science and work is not yet advanced enough.
In the list of data elements of item 4.2 one data element for the UDC and another one for indexing terms or descriptors can be found. It is hoped that the Broad System of Ordering (BSO) which is being developed by UNISIST can be applied later.

4.5 Mandatory and optional data elements

Since the needs of the various organizations co-operating within TermNet will be different, these data elements which are absolutely necessary to describe and identify concepts, documents, organizations, experts, projects, etc. will have to be prescribed within TermNet. Thus two categories of data elements will exist, namely mandatory and optional data elements. A decision on the mandatory data elements to be used within TermNet will have to be taken by the TermNet participants.

REFERENCES

[1] Unesco. Report on the meeting for the evaluation of the International Information Centre for Terminology (Infoterm). UNISIST Steering Committee, 5th meeting. Paris: Unesco, August 1976, 45 p. SC/UNISIST/ST.COM./BUR.5/3.

[2] INFOTERM. First Infoterm Symposium. International Co-operation in Terminology / Premier Symposium d'Infoterm. Coopération internationale en Terminologie. München: Verlag Dokumentation 1976, 332 p., A5. Infoterm Series 3.

[3] FELBER, H. Study on the development of a network for terminology information and documentation. Vienna, 1978, 45 p.

[4] Unesco. Report on the meeting for the evaluation of the Study on the development of a network for terminology information and documentation, prepared by Infoterm. Vienna: Infoterm, Oct. 1977, 12 p.

[5] KROMMER-BENZ, M. Co-operation in TermNet. Analysis of Questionnaires. Vienna, 1978 (see item 11.1).

[6] FELBER, H.; KROMMER-BENZ, M.; MANU, A. International Bibliography of Standardized Vocabularies / Bibliographie internationale de vocabulaires normalisés / Internationale Bibliographie der Normwörterbücher. München: Verlag Dokumentation, 1979, 540 p., A4. Infoterm Series 2.

[7] KROMMER-BENZ, M. World Guide to Terminological Activities / Guide mondial des activités terminologiques. München: Verlag Dokumentation, 1977, 311 p., A5. Infoterm Series 4.

Infoterm
Vienna, Austria

10.3 TERMNET FOLLOW-UP 2

During the past year, i.e. since distribution of our first TermNet follow-up report in November 1978 quite a number of projects and activities could be accomplished, initiated or continued.

1 Infoterm Advisory Board Meeting, Vienna, April 4th, 1979

As indicated in TermNet follow-up No. 1 a meeting of this body was convened spring 1979 to discuss and recommend steps leading to TermNet's actual implementation. Analysis of responses to the inquiry concerning the establishment of TermNet (Doc. TermNet 3, oct. 1978) was submitted to this body. Meanwhile, additional members have joined TermNet (see item 11.1 Analysis of Questionnaires ....)

The Advisory Board recommended the full and immediate implementation of TermNet.

2 TermNet activities

Within the three TermNet programmes approved by this board the following projects are described below:

2.1 TermNet programme I "Developing the scientific basis for terminology", i.e. the General theory and principles of terminology

Recognizing the importance of further theoretical work in terminology Infoterm (in collaboration with other TermNet partners) has already taken the following initiatives and measures:

2.1.1 Adaptation and publication of Wüster's lecture held at the University of Vienna (1973/1974), with the title "Einführung in die Allgemeine Terminologielehre und terminologische Lexikographie" [The general theory of terminology and terminological lexicography - Introduction],

1979 (see item publications). Translations will be prepared in E/F/Pl/Hu/He.

2.1.2 Preparation (and planned publication) of a series with the title "Selected readings in terminology / Textes choisis de terminologie", edited jointly by G. Rondeau (Université Laval, Québec) and H. Felber (Infoterm, Vienna) and published in six volumes by the "Groupe interdisciplinaire de recherche scientifique et appliquée en terminologie" (GIRSTERM) and the Université Laval in Québec (Canada). TermNet partners will be notified of this publication as soon as the first volume is available.

2.1.3 Preparation of the publication of the Proceedings of the "International Symposium on Theoretical and Methodological Problems of Terminology", held in Moscow, November 20th - 23rd, 1979. This Symposium was organized jointly by the Secretariat of ISO/TC 37 WG 1 "Principles of Terminology", the Academy of Sciences of the USSR, Infoterm and AILA.

2.1.4 Publication of a memorial volume for Prof. Dr. Eugen Wüster, with the title "Terminologie als angewandte Sprachwissenschaft" (Terminology - applied linguistics). This volume appeared in 1979 with K.G. Saur Verlag Dokumentation, München - New York - London - Paris, and comprises vol. 1, XIII + 145 p. and vol. 2, 70 p.

2.1.5 Provision for further research into Wüster's numerous projects still left incomplete but of eminent importance to the future development of the theory of terminology. A concrete project dealing with this matter is being supported by the Austrian Foundation for the promotion of scientific research.

2.1.6 Co-editing of a new international journal "Language for Special Purposes (LSP). Research. Teaching. Terminology". This quarterly journal informs about the present state of LSP, theory and teaching, as well as terminology work and research.

2.2 TermNet programme II "Establishing closer co-operation in preparing terminologies"

2.2.1 Terminologies of many subject fields and numerous languages have to be updated constantly to reflect the most recent state of development. This necessitates close collaboration between scientific and professional associations; in addition, modern means of terminological data banks allow rapid updating. Infoterm, therefore, has been and is now particularly involved in establishing closer links between subject specialists elaborating terminologies, information specialists and terminologists involved in the development of international terminological principles to ensure a certain degree of uniformity.

2.2.2 On the occasion of its 70th Anniversary the Union of International Associations (UIA) will organize a World Forum of international and transnational associations (Brussels, June 23 - 27, 1980). One of the proposed working commissions (Commission III) will be dealing with problems of "Transnational communication in the world today" with particular emphasis on language, information and documentation. Infoterm will use this opportunity to raise specific questions related to terminology.

2.3 TermNet programme III "Establishing closer co-operation in collecting, recording, processing and disseminating terminological data and information"

2.3.1 First International Conference on Terminological Data Banks.

2.3.2 BT 1 "International bibliography of terminological articles and books" - progress report.
Bibliographic data on more than 900 articles and other works pertaining to terminology have been collected, classified

and stored in the computer at the "Centre de traitement de l'information" of the University of Laval in Québec, Canada, with the aid of the Bureau des traductions du gouvernement du Canada. The data are now available on tape; final publication is pending.

2.3.3 BT 2 "International Bibliography of Standardized Vocabularies / Bibliographie internationale de vocabulaires normalisés / Internationale Bibliographie der Normwörterbücher"
This bibliography, published in April 1979, covers bibliographical references on some 9 000 terminological standards (relating to any subject field and language treated). It is at present being kept up-to-date with all data stored on tape; publication of supplements is intended. Furthermore, dissemination of these data via different information systems is also being investigated.

2.3.4 BT 3/4 "International Bibliography of multilingual vocabularies in the field of building and construction"
As described in Follow-up No. 1 a special bibliography covering multilingual vocabularies (in some 15 European languages) has been undertaken jointly by Infoterm and the ECE. So far, bibliographical references on approximately 1 000 vocabularies have been assembled and are to be published by 1981.

2.3.5 "Guidelines for recording of bibliographical data"
International bibliographies require unified recording of bibliographical data. For this reason Infoterm prepared draft Guidelines for discussion at the Infoterm Advisory Board meeting; they are now in the process of revision at UNIBID (UNISIST International Centre for Bibliographic Descriptions) to ensure compatibility within UNISIST.

2.3.6 "Guidelines for recording data on organizations, commissions, committees, data banks, etc. involved in terminology work"
These Draft Guidelines also prepared for the meeting mentioned above are being examined with reference to computer application (also with a view to computerization of the "World Guide to terminological activities").

2.3.7 "Vocabulary of terminological data banks"
On the basis of preparatory work carried out by the European Communities, Infoterm and Siemens (Munich) this project will now be taken up full speed in collaboration with all those interested in terminological data banks.

2.3.8 "Comparative study of data elements for terminological data banks"
This study is to be undertaken shortly jointly with all those willing to participate. It is intended to convene a meeting of a working group in 1980.

2.3.9 "A model terminology file"
Such a model for a specific subject field is to be undertaken in the very near future.

2.3.10 Problems relating specifically to data banks storing standardized terminologies are to be dealt with by TermNet in collaboration with the International Organization for Standardization (ISO) and the framework of ISONET. A paper to this effect was presented at the ISO/INFCO meeting, August 1979.

2.3.11 Members of the Infoterm Advisory Board were of the opinion that a TermNet Newsletter represented the first and most important link between TermNet partners. Infoterm intends to start this Newsletter in 1980 with the purpose to inform in greater detail about activities, holdings, services, etc. of other TermNet partners.

An official invitation to supply adequate information will be distributed in due course. Infoterm should welcome any suggestions you might have as regards the layout, content, etc.

11 List of Documents and Publications

List of documents on Infoterm activities and terminology work

E = English
F = French
D = German

1 Infoterm Series

E ☐ Leaflet Infoterm Series 1 "Road to Infoterm"

E/F/D ☐ Leaflet Infoterm Series 2 "International Bibliography of Standardized Vocabularies / Bibliographie internationale de vocabulaires normalisés / Internationale Bibliographie der Normwörterbücher

E/F ☐ Leaflet Infoterm Series 3 "International Co-operation in Terminology / Coopération internationale en Terminologie"

E/F ☐ Leaflet Infoterm Series 4 "World Guide to terminological activities / Guide mondial des activités terminologiques"

E ☐ Leaflet Infoterm

2 Infoterm Newsletters

E ☐ Infoterm Newsletter

3 Reports

E ☐ Unesco. Report on the meeting for the evaluation of the International Information Centre for Terminology (Infoterm). UNISIST Steering Committee, 5th meeting, 45 p.

E ☐ Unesco. Report on the meeting for the evaluation of the study on the development of a network for terminology information and documentation, prepared by Infoterm, 12 p.

4 TermNet

E ☐ Infoterm. Report on the First International Conference on Terminological Data Banks, 18 p. (TermNet 4(9)-79)

E ☐ Infoterm. Report on the Infoterm Advisory Board Meeting, 9 p. (TermNet 5(6)-79)

E ☐ FELBER, H. Infoterm and TermNet activities (Diagram and description) (Infoterm 4-80)

E ☐ FELBER, H. Study on the development of a network for terminology information and documentation, 46 p. (Infoterm 1-78)

E ☐ KROMMER-BENZ, M. Co-operation in TermNet. Analysis of Questionnaires, 21 p. (TermNet 3-78)

E ☐ KROMMER-BENZ, M. Infoterm and TermNet. International Activities in terminology, 19 p.

E ☐ KROMMER-BENZ, M. TermNet and INTERCONCEPT, 17 p. (Infoterm 12-78)

E ☐ FELBER, H.; KROMMER-BENZ, M. TermNet follow-up-activities, 4 p., (TermNet 6(1)-79).

E ☐ FELBER, H.; KROMMER-BENZ, M. TermNet follow-up 2, 3 p. (TermNet 6(2)-79).

F ☐ MANU, A. Infoterm et le réseau de terminologie TermNet, 9 p. (Infoterm 11-79)

5 Infoterm

D ☐ FELBER, H. Infoterm - Knotenpunkt der internationalen Terminologiearbeit, 17 p. (Infoterm 5-78)

D ☐ FELBER, H.; KROMMER-BENZ, M. Das Internationale Informationszentrum für Terminologie (Infoterm) - ein Brennpunkt der internationalen Terminologiearbeit, 15 p. (Infoterm 6-78)

D ☐ KROMMER-BENZ, M. Das Internationale Informationszentrum für Terminologie (Infoterm) im Österreichischen Normungsinstitut, 4 p.

6 Theory of terminology

E, D ☐ FELBER, H. Theory of terminology, terminology work and terminology documentation, 13 p. (Infoterm 1-79)

E ☐ FELBER, H. Development of the general theory of terminology and of terminology work - A short history, 20 p. (Infoterm 9-78)

E ☐ FELBER, H. The Vienna School of Terminology - fundamentals of its theory, 9 p. (Infoterm 12-79)

F ☐ WÜSTER, E. La théorie générale de la terminologie - un domaine interdisciplinaire impliquant la linguistique, la logique, l'ontologie, l'informatique et les sciences des objets, 10 p.

D ☐ WÜSTER, E. Die Allgemeine Terminologielehre - Ein Grenzgebiet zwischen Sprachwissenschaften, Logik, Ontologie, Informatik und den Sachwissenschaften, 46 p.

D ☐ WÜSTER, E. Einführung in die Allgemeine Terminologielehre und Terminologische Lexikographie, 210 p. (Infoterm 17-79)

D ☐ WÜSTER, E. Begriffs- und Themaklassifikation. Unterschieder in ihrem Wesen und ihrer Anwendung, 8 p.

D ☐ WÜSTER, E. Die Umkehrung einer Begriffsbeziehung und ihre Kennzeichnung in Wörterbüchern, 8 p.

D ☐ FELBER, H. Grundsätze der allgemeinen Terminologielehre und die Terminologienormung, 3 p. (In foterm 14-78)

E ☐ KROMMER-BENZ, M. Infoterm - Activities concerning methodological and theoretical aspects of terminology work, 8 p. (Infoterm 14-79)

□ KROMMER-BENZ, M. Report on the "International Symposium on Theoretical and Methodological Problems of Terminology", 11 p. (Infoterm 15-79)

7 Standardization of terminology

/F/D □ FELBER, H. International efforts to overcome the difficulties in technical communication, 14 p.

□ FELBER, H. International standardization of terminology - Theoretical and methodological aspects, 16 p.

D □ ISO/TC 37 "Terminology". Fifth meeting, 1974, 2 p.

F □ ALMASY, CH. La normalisation internationale en terminologie. 5e Réunion du Comité Technique ISO/TC 37 'Terminologie (principes et coordination)', 6 p.

/F □ FELBER, H. Meeting of ISO/TC 3/WG 1 "Principles of terminology", September 27-30, 1976 in Moscow, 3 p.

□ FELBER, H. Meeting of ISO/TC 37/WG 3 "Layout of vocabularies" from 22 to 24 May 1978 in Paris - Report of TC 37 Secretariat, 4 p. (Infoterm 8-78)

□ KROMMER-BENZ, M. Infoterm - International activities in the area of standardized terminology, 5 p.

□ FELBER, H. Information on standardized vocabularies (terminological standards) and standardized concepts - a joint task of INFCO and Infoterm, 8 p. (Infoterm 8-79)

D □ FELBER, H. Die internationale terminologische Grundsatznormung - Rückblick und Ausblick, 11 p. (Infoterm 7-78

8 Terminology work and documentation

D □ FELBER, H. Theory of terminology, terminology work and terminology documentation, 13 p. (Infoterm 1-79)

□ FELBER, H. The development of national and international terminology work and documentation, 21 p. (Infoterm 11-78)

D □ FELBER, H. Organization of terminology work at international level (Infoterm 10-78)

□ FELBER, H. The specific needs of data exchange in terminology. Bibliographic and terminological data elements, 20 p.

D □ WÜSTER, E. Die vier Dimensionen der Terminologiearbeit und die Internationale terminologische Grundsatzarbeit, 12 p.

D □ KROMMER-BENZ, M. Stand und Entwicklung der internationalen Terminologiearbeit und ihre Bedeutung für die Entwicklung von Informationssystemen, 11 p.

D □ FELBER, H. Erfassungsschemata zur Begriffsdokumentation, 8 p. (Infoterm 4-78)

D □ FELBER, H. Terminologieprobleme im Informations- und Dokumentationswesen - Das terminologische Fachwörterbuch und der Thesaurus, 9 p. (Infoterm 4-79)

D ☐ MANU, A. Wesen und Funktionen der Terminologiedokumentation, 18 p. (Infoterm 16-78)

D ☐ KROMMER-BENZ, M. Die internationale Zusammenarbeit auf dem Gebiet der Terminologie, 7 p. (Infoterm 13-78)

D ☐ FELBER, H. Die moderne Terminologiearbeit - das terminologische Wörterbuch und der Dokumentationsthesaurus, 23 p. (Infoterm 13-79)

D ☐ MANU, A. Internationale Bibliographie mehrsprachiger Wörterbücher auf dem Gebiet des Bauwesens (Infoterm 2-80)

PUBLICATIONS

COMMISSION ELECTROTECHNIQUE INTERNATIONALE / INTERNATIONAL ELECTROTECHNICAL COMMISSION. Vocabulaire Electrotechnique International / International Electrotechnical Vocabulary. Index General / General Index. Edition provisoire / Provisional edition. Clamart: CEI/IEC, 1979, 1011 p., 18x26.

GOSUDARSTVENNYJ KOMITET STANDARTOV SOVETA MINISTROV SSSR. Standartizacija terminologii v SSSR i meždunarodnyh organizacijah [Standardization of terminology in USSR and international organizations]. Moskva: Gosudarstvennyj komitet standartov soveta ministrov SSSR. 1978, 48 p., A5.

FELBER, H; KROMMER-BENZ, M; MANU, A. International bibliography of standardized vocabularies / Bibliographie internationale de vocabulaires normalisés / Internationale Bibliographie der Normwörterbücher. München: K.G. Saur, 1979, 2nd edition, 540 p., A4 (Infoterm Series 2).

GESELLSCHAFT FÜR KLASSIFIKATION E.V. Kooperation in der Klassifikation I. / Co-operation in classification I. Proceedings der Sekt. 1-3 der 2. Fachtagung der Gesellschaft für Klassifikation e.V. Frankfurt-Höchst, 6.-7. April 1978, Frankfurt: Gesellschaft für Klassifikation e.V., 1978, 211 p., A5. (SK 2.)

GESELLSCHAFT FÜR KLASSIFIKATION E.V. Kooperation in der Klassifikation II / Co-operation in classification II. Proceedings der Sekt. 4-6 der 2. Fachtagung der Gesellschaft für Klassifikation e.V. Frankfurt-Höchst, 6.-7. April 1978, Frankfurt: Gesellschaft für Klassifikation e.V., 1978, 229 p., A5. (SK 3.)

WÜSTER, E. Einführung in die allgemeine Terminologielehre und Terminologische Lexikographie General Theory of Terminology and Terminological Lexicography. - An introduction . 2 parts. Bauer, L. (ed.), Wien/New York: Springer Verlag, 1979, XIII+145 and 70 p., A5.

FELBER, H; LANG, F; WERSIG, G. Terminologie als angewandte Sprachwissenschaft. Gedenkschrift für Univ.-Prof. Dr. Eugen Wüster. München/New York/London/Paris: K.G. Saur, 1979, 272 p., A5.

Infoterm Series

Published for the International Centre for Terminology, Vienna

Band 1
The Road to Infoterm
Herausgegeben von
Eugen Wüster
1974. XII, 144 Seiten. Br. DM 36,–
ISBN 3-7940-5501-2

Band 2
International Bibliography of Standardized Vocabularies
Begründet von Eugen Wüster
Bearbeitet von Helmut Felber, Magdalena Krommer-Benz und Adrian Manu
1979. 2. Ausgabe. XXIV, 540 Seiten. Lin. DM 148,–
ISBN 3-598-05502-1

Band 3
International Cooperation in Terminology
First Infoterm Symposium, Vienna 1975
1976. XII, 333 Seiten. Br. DM 48,–
ISBN 3-7940-5503-9

Band 4
World Guide to Terminological Activities, Organisations, Terminology Banks, Committees
Bearbeitet von
Magdalena Krommer-Benz
1977. 311 Seiten. Br. DM 68,–
ISBN 3-7940-5504-7

Band 5
Terminological Data Banks
Proceedings of the First International Conference convened Vienna 2 and 3 April 1979 by Infoterm
Herausgegeben von
Christian Galinski
1980. 207 Seiten. Br. ca. DM 48,–
ISBN 3-598-21365-4

K·G·Saur
München · New York · London · Paris
K·G·Saur Verlag KG, Postfach 711009, 8000 München 71
Tel. (089) 798901. Telex 05212067 saur d

Bei Fragen zur Produktsicherheit wenden Sie sich bitte an:
If you have any questions regarding product safety,
please contact:

Walter de Gruyter GmbH
Genthiner Straße 13
10785 Berlin
productsafety@degruyterbrill.com